Advances in Industrial Control

Springer

London
Berlin
Heidelberg
New York
Barcelona
Budapest
Hong Kong
Milan
Paris
Santa Clara
Singapore
Tokyo

Rami S. Mangoubi

Robust Estimation and Failure Detection

A Concise Treatment

With 49 Figures

 Springer

Dr Rami S. Mangoubi
Draper Laboratories, 555 Technology Square, Cambridge,
Massachusetts 02139-3563, USA

ISBN-13: 978-1-4471-1588-5   e-ISBN-13: 978-1-4471-1586-1
DOI: 10.1007/978-1-4471-1586-1

British Library Cataloguing in Publication Data
Mangoubi, Rami S.
   Robust estimation and failure detection
   : a concise treatment. - (Advances in industrial control)
   1.Linear systems 2.Estimation theory 3.Control theory
   4.Kalman filters
   I.Title
   629.8'32
   ISBN-13: 978-1-4471-1588-5

Library of Congress Cataloging-in-Publication Data
Mangoubi, Rami.
     Robust estimation and failure detection : a
   concise treatment / Rami S. Mangoubi.
        p.   cm. -- (Advances in industrial control)
     ISBN-13: 978-1-4471-1588-5
     1. Linear control systems.   I.   Title.   II. Series.
   TJ220.M36   1998
   629.8'32--dc21   98-4616

© Springer-Verlag London Limited 1998
Softcover reprint of the hardcover 1st edition   1998

Typesetting: Camera ready by author

69/3830-543210 Printed on acid-free paper

Advances in Industrial Control

Series Editors

Professor Michael J. Grimble, Professor of Industrial Systems and Director
Dr. Michael A. Johnson, Reader in Control Systems and Deputy Director

Industrial Control Centre
Department of Electronic and Electrical Engineering
University of Strathclyde
Graham Hills Building
50 George Street
Glasgow G1 1QE
United Kingdom

Series Advisory Board

Professor Dr-Ing J. Ackermann
DLR Institut für Robotik und Systemdynamik
Postfach 1116
D82230 Weßling
Germany

Professor I.D. Landau
Laboratoire d'Automatique de Grenoble
ENSIEG, BP 46
38402 Saint Martin d'Heres
France

Dr D.C. McFarlane
Department of Engineering
University of Cambridge
Cambridge CB2 1QJ
United Kingdom

Professor B. Wittenmark
Department of Automatic Control
Lund Institute of Technology
PO Box 118
S-221 00 Lund
Sweden

Professor D.W. Clarke
Department of Engineering Science
University of Oxford
Parks Road
Oxford OX1 3PJ
United Kingdom

Professor Dr -Ing M. Thoma
Westermannweg 7
D-30419 Hannover
Germany

Professor H. Kimura
Department of Mathematical Engineering and Information Physics
Faculty of Engineering
The University of Tokyo
7-3-1 Hongo
Bunkyo Ku
Tokyo 113
Japan

Professor A.J. Laub
College of Engineering - Dean's Office
University of California
One Shields Avenue
Davis
California 95616-5294
United States of America

Professor J.B. Moore
Department of Systems Engineering
The Australian National University
Research School of Physical Sciences
GPO Box 4
Canberra
ACT 2601
Australia

Dr M.K. Masten
Texas Instruments
2309 Northcrest
Plano
TX 75075
United States of America

Professor Ton Backx
AspenTech Europe B.V.
De Waal 32
NL-5684 PH Best
The Netherlands

SERIES EDITORS' FOREWORD

The series *Advances in Industrial Control* aims to report and encourage technology transfer in control engineering. The rapid development of control technology impacts all areas of the control discipline. New theory, new controllers, actuators, sensors, new industrial processes, computer methods, new applications, new philosophies..., new challenges. Much of this development work resides in industrial reports, feasibility study papers and the reports of advanced collaborative projects. The series offers an opportunity for researchers to present an extended exposition of such new work in all aspects of industrial control for wider and rapid dissemination.

This monograph brings together two of the most exciting areas of advanced signal processing and control. The first involves the development of optimal estimators for uncertain signal and noise models. The second is the subject of failure detection and isolation, which has considerable potential in a range of applications. The text provides a gradual build-up of ideas moving from traditional Wiener and Kalman filtering to risk sensitive control and estimation problems.

The solution of a robust estimation problem is crucial to solving the fault monitoring and detection problem effectively. That is, one of the main difficulties is deciding whether an estimated fault condition is really due to a fault or is simply due to poor system models. A robust estimation method is therefore essential. The text provides a novel solution of a class of robust estimation problems and these are in a very suitable form for application to the failure detection and isolation problem considered in the later chapters. It also includes convincing application studies for both marine and aerospace applications. There are also extensive appendices covering related and relevant areas.

The Monograph should be a valuable addition to the literature on model based fault monitoring and detection methods and is a timely contribution since interest in this subject is intense.

M.J. Grimble and M.A. Johnson
Industrial Control Centre
Glasgow, Scotland, UK

PREFACE

This monograph, which is an outgrowth of my doctoral thesis, is a concise treatment of recursive estimation and failure detection for dynamic plants. In my view, the monograph possesses two unique features. First, it treats both optimal or Kalman filtering, *as well as* the relatively new subject of robust filtering, or filtering for plants with uncertain dynamics and disturbance models. Second, it combines the two subjects of state estimation *and* failure detection for dynamic systems into one entity. It is noteworthy that, in the signal processing literature, the twin subjects of estimation and detection are treated together. A monograph that does the same for dynamic systems is, in my view, a useful addition to the literature. The treatment of failure detection includes algorithms that make use of the Kalman filter, such as the generalized likelihood ratio test, and newer algorithms that rely on robust filters.

In writing this monograph, I had several audiences in mind. One such audience is the body of students, researchers, and practicing engineers who work with, or whose interest include, recursive estimation for dynamic systems. The last fifteen years have seen considerable progress in the theory and implementation of control systems with robustness to both disturbance and plant model uncertainties. As a result, compensators that achieve both robust stability and performance objectives are not uncommon, at least for linear plants. These advances in control can benefit filtering theory and design, and I have made every attempt to develop the robust filtering methodology in a way that the reader with no background in control theory can fully appreciate. All the new filters derived are extensions of the Kalman, or H_2, filter, and they all reduce to it when disturbance and plant uncertainties vanish.

Another audience is engineers and researchers interested in the problems of plant health monitoring and failure diagnosis. The design of residual generators that are insensitive to uncertainties and rapidly varying dynamics, but at the same time are sensitive to failures, is a nontrivial task. The approach suggested here is a heuristic extension of the generalized likelihood ratio test (GLRT) for failure detection and isolation that makes use of robust filters.

The reader who is new to both filtering and detection theory has many introductory books to choose from. The material presented here is certainly accessible to such a reader, though it is different from other books in several ways. First, the overview chapter gives a summary of the Wiener and Kalman filters. An appendix contains a tutorial on these filters, including derivations, as well as other topics. The Kalman filter's deterministic and stochastic properties are listed in the overview. More important, the limitations of this filter are discussed and demonstrated at an early stage. A formulation for the problem of failure detection is also presented in the overview chapter and the role of the Kalman filter in this approach to detection is quickly seen. Furthermore, the filters derived in this monograph are more general than the Kalman filter, and by setting one or more parameters in these filters to a particular value, the Kalman filter is recovered. Smoothing and the extended Kalman filter (EKF) are also part of the appendix on Kalman filters.

It is also my hope that the monograph will encourage more researchers in the areas of estimation and detection to be curious about the latest progress in control, so that new and better methodologies can be exported into these two fields. Equipped with the background given by this monograph, as well as the basics of linear systems theory, articles and books on robust control become more easily accessible, or perhaps much less mysterious. This, at least, has been my experience.

Likewise, control engineers with no prior background or interest in the problems of estimation and detection will see the impact that developments in their field have on these two areas of research.

The subject matter requires some familiarity with linear systems and probability. This is generally assumed of advanced undergraduate and graduate level students. There is some benefit in proceeding through the chapters in sequence, but this is not essential, especially if the reader has a background in either Kalman filtering or robust control. Chapter 2, and Appendix A if needed, cover the Kalman filter and the generalized likelihood ratio test (GLRT) for failure detection and isolation. Chapter 2 also introduces the small gain theorem and risk sensitivity. Chapter 3 treats the subject of game theoretic and robust filtering for discrete-time systems, with a summary of the continuous-time case left to Appendix C. The formulation in Chapter 3 and Appendix C is deterministic. Chapter 4 presents a stochastic interpretation of robust filtering - risk sensitive estimation - of which minimum error variance estimation, or Kalman filtering, is a special case. This short chapter makes the connection between risk sensitivity and robust filtering. Chapter 5 deals with robust detection and isolation. Two applications are given in Chapter 6. The first application deals with robust failure detection for an unmanned underwater vehicle, and the second is a problem in both robust estimation and failure detection for reentry vehicles.

Draper Laboratory, which was the first laboratory to use Kalman filters in the early sixties (for the Apollo navigation system), was also the first to encourage the development of robust estimation theory. It is my pleasure to acknowledge the help given me by Draper Laboratory. In particular, Neil Adams and Jim Negro supported the writing of this monograph. Neil never stopped encouraging me. Byong Ahn provided some financial support. I am fortunate to have as close friends and colleagues Roger Hain and Naz Bedrossian. Roger shared his students with me, and was always generous with his time. Naz, and his wife Melinda, made sure my spirit was up and kept me going. George Verghese, Brent Appleby, Wallace VanderVelde, and Milton Adams, all members of my doctoral committee, made invaluable comments as I was working on my thesis, and their high standards will always be an inspiration. My M.I.T. graduate student Christian Jacquemont provided excellent feedback, and his skepticism kept me honest. Special thanks go to M.I.T. graduate students and Draper Laboratory Fellows Ramses Agustin, Chris Dever, and Nhut Ho, all of whom I supervised together with Roger Hain. Ramses worked on one of the applications in the monograph, while Chris and Nhut read the manuscript, and provided excellent feedback. Keith Rogers and Mark McConley took time from their busy schedule to read parts of the manuscript. Manuel Martinez-Sanchez made excellent suggestions on the Kalman filter appendix. Draper lawyer Sharon Shore helped me with the paper work, while Anne Neill and Nick Pinfield of Springer Verlag were very patient and supportive. Special thanks are due to Michael Grimble for offering me this rare opportunity to turn my thesis into a monograph. Michael

Johnson made good suggestions regarding the application chapter. My entire family, especially my wife and best friend (Uschi, si je ne t'avais pas rencontrée, j'aurais fait le tour du monde pour te chercher) has been a source of love and support without which I cannot live. Our children Oren, Tomer, and Daniel have been a source of inspiration.

Cambridge, Massachusetts

CONTENTS

Contents

CHAPTER 1
INTRODUCTION

Generally speaking, the subjects of estimation and failure detection for dynamic systems are concerned with extracting information of interest about a dynamic plant or process. The objective is to estimate physical quantities related to the plant, as well as to determine whether any abrupt change or failure occured. The approach consists of processing measurements obtained from sensors and combining it with a priori information available from the plant's dynamic model. One possible application is the monitoring of dynamic plants, such as vehicles, in order to estimate their position and velocity, and to detect plant failures. Another application is target tracking, where the objective is to follow targets and detect their maneuvers.

The purpose of this book is to provide a concise treatment of classical and current estimation and failure detection theory. Specifically, the book starts with the first linear least squares, or minimum variance, filter developed in the 1940's, the Wiener filter. This filter is applicable to linear time invariant systems at steady state, and its solution is based in the frequency domain. Next, we introduce the Kalman filter, which is a recursive time domain filter. This filter is shown to be a generalization of the Wiener filter. It is applicable to linear time varying and transient systems, and is equivalent to the Wiener filter for linear time invariant plants at steady state. These two classical filters, in particular the Kalman filter, which is easily implementable on digital computers, have seen pervasive use in many applications. Their role in failure detection and isolation is illustrated. Specifically, we derive the generalized likelihood ratio test (GLRT), which is shown to require the Kalman filter for its implementation. The GLRT is an illustration of how estimation and detection are tightly coupled subjects.

The Kalman and Wiener filters, as well as the likelihood ratio test, assume perfect knowledge of the plant dynamics and noise statistics. We first expand the theory by dealing with noise model uncertainty. In fact, we derive a family of recursive estimators, of which the Kalman and Wiener filters are members. These filters can trade off optimal minimum variance, or least squares error, performance against noise robustness. We discuss this generalization in both the deterministic and stochastic context, and later, illustrate how the new analytical framework relates to the likelihood ratio tests.

Finally, we develop the analytical basis for dealing with *both* noise and plant model uncertainties. We apply the theory to derive robust filters. The filters, in turn, are used to develop failure detection and isolation algorithms that are sensitive to failures, but insensitive to the modeling errors. The development of this theory is more complex, and the filters require the solution to two matrix Riccati equations, as opposed to the previous filters, which need the solution to only one Riccati equation. In the absence of uncertainites, the robust filter reduces to the Kalman filter, and the associated robust detection algorithm reduces to the likelihood ratio test.

Chapter 2 begins by reviewing the development of estimation theory for linear systems. Basic results for the Wiener and Kalman filters are summarized in the chapter, while Appendix A provides a concise treatment of these two topics, including derivations and extension to nonlinear systems. The deleterious effects of modeling errors on the Kalman and Wiener filters are demonstrated in Chapter 2. The small gain theorem, which is the foundation of robust estimation, is also introduced.

Chapter 2 also describes the problem of failure detection and isolation in dynamic systems, with special emphasis on the generalized likelihood ratio test, the role of the Kalman filter in this test, and the issue of modeling uncertainties.

In Chapter 3, a game theoretic approach is developed for the solution of the robust estimation problem for discrete-time linear systems. The chapter begins by developing estimators with robustness to noise model uncertainties only. The approach is then extended to the derivation of a more general filter, which is robust to both plant and noise model uncertainties. The robust fixed-interval smoothing problem is also treated. The chapter ends with numerical examples, including a navigation application.

In Appendix C, the robust H_∞ filter for continuous-time linear systems is derived. The derivation is similar to that of Chapter 3 for discrete-time systems. For this reason, the appendix proceeds directly with the derivation of the general filter with robustness to both plant and noise model uncertainties, and does not start with the noise only case.

Chapter 4 is concerned with the stochastic interpretation of robust game theoretic (including H_∞) estimation. Specifically, it is shown that the stochastic risk sensitive estimation problem is similar in nature to deterministic filtering problems treated in earlier chapters. The chapter starts by reviewing the relationship between the two problems for the case where no plant model perturbations are assumed. This relationship is then extended to the case where the plant dynamics are uncertain.

In Chapter 5, the estimators developed in earlier chapters are used for the design of failure detection and isolation algorithms that are robust to noise, plant, and failure mode uncertainty. Here, we extend the relationship between estimation and detection to a larger class of problems.

Chapter 6 presents two applications. The first is concerned with failure detection for an unmanned undersea vehicle (UUV), while the second is a detection and estimation problem for a reusable launch vehicle (RLV).

CHAPTER 2

ESTIMATION AND FAILURE DETECTION: AN OVERVIEW

2.1 Introduction

In this chapter, we present an overview of estimation and failure detection for linear systems. In Section 2.2, we discuss linear filtering and its past and current development. First, Section 2.2.1 gives a brief description of the Wiener and Kalman filters, the most commonly used filters for dynamic systems. Only the basic results are summarized in the section; while the derivation and other topics are treated in Appendix A. Next, in Section 2.2.2, we briefly discuss problems that are more general than linear least squares estimation, including robust filtering, which is one of our main concerns in subsequent chapters. We then motivate the need for robust filters in Section 2.2.3 by demonstrating the deleterious effect of model uncertainties on the Kalman filter. A formulation of the robust filtering problem is given in Section 2.2.4. The small gain theorem, which is the key to solving that problem, is also introduced in that section. In Section 2.2.5, we describe how the small gain theorem is used in robust estimation. The actual derivation of the filters is left to Chapter 3. Further discussion on robust estimation is found in Section 2.2.6. The robust filtering problem is formulated in Section 2.2.4 in a deterministic context. In Section 2.2.7, we introduce the risk sensitive estimation problem, a stochastic version of robust filtering.

Section 2.3 offers a similar discussion for the problem of failure detection and isolation (FDI). Emphasis is on the role of the Kalman filter. In Sections 2.3.1 and 2.3.2, we introduce the generalized likelihood ratio test for two different classes of failures, additive and non-additive, respectively. The sensitivity of likelihood ratio tests to modeling uncertainties is demonstrated in Section 2.3.3, motivating the subject of robust detection, an overview of which is given in Section 2.3.4. The robust detection methodology suggested in this book, which makes use of robust filters as residual generators, is the subject of Chapter 5.

2.2   Ever Since Wiener

The first model based estimator for linear systems, the Wiener filter, was developed by Norbert Wiener [113] in the 1940's during the Second World War in order to track fighter planes. Its equivalent in the state space domain, the Kalman-Bucy filter [63], [64], first appeared about two decades later in the late 1950's. Since this period, the use of these filters has not only been pervasive in areas of engineering such as target tracking and aircraft control, but has also found applications in a wide variety of fields such as image processing [95], [118], metallurgy [98], geophysics [48], [82], environmental sciences [41], and cattle production [98]. This widespread use occurs despite the fact that both the Wiener and Kalman filters rely on having an accurate model of the system whose state they estimate, an assumption that is questionable in most practical contexts.

It should be noted that linear filters or observers other than the Wiener and Kalman filter also exist. These include general purpose filters, such as the Luenberger observer [78], or more specialized ones, such as the detection filter of Beard [16] and Jones [59], which is used for failure detection and isolation. These filters are also sensitive to model uncertainty. The background discussion of the next section will concentrate on the Wiener and Kalman filters, but similar discussions could be developed for other filters.

Before we discuss the effect of modeling uncertainties on the behavior of these filters, we will give a description of the two filters. Our goal is to state the formulation of the minimum error variance or least squares estimation problem for linear dynamic systems, to discuss the filters' properties, give their equations and, more importantly, to point to the restrictions on the use of each filter that motivated further developments in the field. The reader interested in a derivation and a

more detailed treatment of the Kalman and Wiener filters is referred
to Appendix A, which requires no previous knowledge of the subject.
The use of the Kalman filter has been extended to nonlinear systems.
The extended Kalman filter (EKF), which is an approximate nonlinear
filter, is also treated in Appendix A. A detailed historical survey on
linear filtering is given by Kailath in [61].

2.2.1 Wiener and Kalman Filters

Consider two jointly wide-sense stationary random processes $x(t)$ and
$y(t)$ with *known* auto- and cross-correlation functions given, for all t,
by

$$
\begin{aligned}
R_{xx}(t) &\equiv E\left(x(s)x(s+t)\right) \\
R_{yy}(t) &\equiv E\left(y(s)y(s+t)\right) \\
R_{xy}(t) &\equiv E\left(x(s)y(s+t)\right)
\end{aligned}
$$

where $E(\cdot)$ is the expectation operation. The mathematical problem,
formulated by E. Hopf, and addressed by N. Wiener at the M.I.T.
Radiation Laboratory in the 1940's for application to antiaircraft control systems is the following: Given measurements $\{y(\tau), -\infty < \tau < t\}$,
find the linear estimate $\hat{x}(t)$ of $x(t)$ that minimizes the expected value
of the squared error at any time t. More formally,

$$
\min_{\hat{x}} \quad E\left(x(t) - \hat{x}(t)\right)^2 , \quad -\infty < t < \infty \tag{2.1}
$$

$$
\text{with} \quad \hat{x}(t) = \int_{-\infty}^{t} h(t,\tau)y(\tau)d\tau \tag{2.2}
$$

The function $h(t,\tau)$ is the impulse response of a causal linear system,
i.e., $h(t,\tau) = 0, t < \tau$. It can be shown that the assumption of joint
stationarity of the processes x and y and the semi-infinite observation
window together imply that the optimal filter is a linear time-invariant
one, i.e., (with the usual abuse of notation),

$$
h(t,\tau) = h(t-\tau)
$$

The optimal filter satisfies the Wiener-Hopf equation, given by

$$
R_{xy}(t) = \int_{-\infty}^{\infty} h(t-\tau)R_{yy}(\tau)d\tau \quad , \quad t > 0 \tag{2.3}
$$

Note that the presence of the constraint $t > 0$ in the above equation prevents us from taking the bilateral Laplace or Fourier transform

of the equation. A more sophisticated technique is therefore needed, and such a technique was developed by Wiener. If $\mathcal{H}_W(s)$, $S_{xx}(s)$, $S_{xy}(s)$, and $S_{yy}(s)$ are the Laplace transform of, respectively, $h(t)$, $R_{xx}(t)$, $R_{xy}(t)$, and $R_{yy}(t)$, then the solution to the Wiener-Hopf equation can be given in terms of these transforms

$$\mathcal{H}_W(s) = \frac{1}{S_{yy}^+(s)} \left\{ \frac{S_{xy}(s)}{S_{yy}^-(s)} \right\}^+ \tag{2.4}$$

where the superscripts $^+$ and $^-$ denote respectively, the causal and anti-causal portion of the transform.

In discrete time[1], the Wiener filtering problem is similar: Given two wide-sense jointly stationary processes x_k and y_k with known auto- and cross-correlation functions, the goal is to solve the optimization problem

$$\min_{\hat{x}} \quad E\left(x_k - \hat{x}_k\right)^2 , \quad -\infty < k < +\infty \tag{2.5}$$

$$\text{with} \quad \hat{x}_k = \sum_{\ell=-\infty}^{k-1} h[k,\ell] y_\ell \tag{2.6}$$

The optimal filter, which is time invariant, must satisfy the discrete Wiener-Hopf equation [88]

$$R_{xy}[k] = \sum_{\ell=-\infty}^{\infty} h[k-\ell] R_{yy}[\ell] , \quad k \geq 0 \tag{2.7}$$

where

$$
\begin{aligned}
R_{xx}[k] &\equiv E\left(x_\ell x_{\ell+k}\right) \\
R_{yy}[k] &\equiv E\left(y_\ell y_{\ell+k}\right) \\
R_{xy}[k] &\equiv E\left(x_\ell y_{\ell+k}\right)
\end{aligned}
$$

and the solution is given in the z-transform domain by the expression

$$\mathcal{H}_W(z) = \frac{1}{S_{yy}^+(z)} \left\{ \frac{S_{xy}(z)}{S_{yy}^-(z)} \right\}^+ \tag{2.8}$$

Eqs.(2.3) and (2.7) have their restrictions. These include:

---

[1]We deal almost entirely with discrete-time systems. The only exceptions are portions of this section, where we describe the original problem that Wiener and Hopf solved, and Appendix C, which presents a summary of continuous-time robust estimation. One reason for this preference is that the mathematics of discrete-time filtering are easier, at least when the setting is stochastic. Another reason is that failure detection algorithms make use of transient filters, which are easier to implement in discrete time.

1. semi-infinite observation intervals,

2. jointly wide-sense stationary signal and observation processes,

3. scalar processes, and

4. accurate knowledge of the auto- and cross-correlation functions.

Efforts to remove the first three restrictions were under way soon after Wiener's work was understood. Robust filtering, one of the subjects of this book, addresses the fourth restriction.

To remove the first three restrictions, Eq.(2.2) can be replaced by

$$\hat{x}(t) = \int_0^T h(t,\tau)y(\tau)d\tau \tag{2.9}$$

The limit on the integral removes the first restriction mentioned above, while the two arguments t and τ make the impulse response time-varying, thus removing the second restriction. The third restriction is easily removed if $h(t,\tau)$ is a matrix function, and $y(t)$ and $\hat{x}(t)$ are vector processes. Three possibilities exist:

1. If $T < t$, then $\hat{x}(t)$ is a *predicted* estimate, as observations are not available beyond time T.

2. If $T = t$, then $\hat{x}(t)$ is a *filtered* estimate.

3. If $T > t$, then $\hat{x}(t)$ is a *smoothed* estimate.

The impulse response $h(t,\tau)$ must now satisfy

$$R_{xy}(t,s) = \int_0^T h(t,\tau)R_{yy}(\tau,s)d\tau, \quad 0 \le s \le T \tag{2.10}$$

which is more general than Eq.(2.3). If the observation process has an additive white noise component $v(t)$ with intensity $R(t)$ that is uncorrelated with $x(t)$, then its correlation function can be written as

$$R_{yy}(\tau,s) = R(\tau)\delta(\tau - s) + K_{yy}(\tau,s)$$

Eq.(2.10) then becomes

$$R_{xy}(t,s) = R(s)h(t,s) + \int_0^T h(t,\tau)K_{yy}(\tau,s)d\tau, \quad 0 \le s \le T \tag{2.11}$$

Likewise, in discrete time, the analog of Eq.(2.9) is

$$\hat{x}_k = \sum_{\ell=0}^{K} h[k,\ell] y_\ell$$

where the function h is determined from the discrete analog of Eq.(2.10) [88].

$$R_{xy}[k,s] = \sum_{\ell=0}^{K} h[k,\ell] R_{yy}[\ell,s], \quad 0 \leq s \leq K \tag{2.12}$$

The integral equation (2.10) is called Fredholm's equation of the first kind, and its solutions must be interpreted as having some singularities at times $t = 0$ and $t = T$. The addition of a white noise component in Eq.(2.11), which is called Fredholm's equation of the second kind, suppresses these singularities. As a result, it is easier to solve. But even in this case, the solution is difficult to compute numerically (e.g., [73], [108], and [110]), especially for vector processes, since it is not recursive. As a result, the solution has to be solved again from scratch if the observation interval is increased. In discrete time, Eq.(2.12) is essentially a set of linear equations whose complexity is on the order of the cube of the observation interval [88], and which must be solved again if the interval changes. All these difficulties came to the fore in the late 1950's in connection with the problem of determining satellite orbits.

To overcome the computational difficulties of Eqs.(2.10), (2.11), and (2.12), effective recursive least squares estimation algorithms were desired. Among the first studies were those of Swerling [105], and Kalman and Bucy [63, 64]. Here, we will be concerned with the Kalman filter in discrete time.

Kalman's work was characterized by a new assumption. Specifically, accurate knowledge of the correlation functions was replaced by accurate knowledge of a finite-dimensional model of the system or physical process generating the observation signal. Specifically, instead of the specifications R_{xx}, R_{xy}, and R_{yy}, we now have a state-space model of the form

$$
\begin{aligned}
x_{k+1} &= A_k x_k + B_k r_k & (2.13) \\
y_k &= C_k x_k + D_k r_k & (2.14) \\
E(r_k r_l') &= \delta_{kl} I & (2.15) \\
E\left(r_k (x_0 - \hat{x}_0)'\right) &= 0 & (2.16) \\
E\left((x_0 - \hat{x}_0)(x_0 - \hat{x}_0)'\right) &= P_0 & (2.17)
\end{aligned}
$$

Here the real vector stochastic process x_k is given by a shaping filter whose input is a zero-mean white or uncorrelated process noise $B_k r_k$, and the observation signal y_k is the output of a sensor, corrupted by measurement noise $D_k r_k$. The matrices A_k, B_k, C_k, and D_k, of appropriate dimensions, represent a physical plant and its sensors, or any system of interest. The initial condition is x_0, while $\hat{x}_0$ is the initial estimate, and P_0 is the positive semi-definite covariance matrix of the initial error $\tilde{x}_0 \equiv x_0 - \hat{x}_0$. It is assumed that the initial error is uncorrelated with the subsequent noise. Finally, in the above formulation, the noise vector r_k is normalized to have unit covariance.

Any desired process or measurement noise covariance, however, can be obtained by adjusting the values of the matrices B_k and D_k. Moreover, if $B_k D_k' = 0$, then the process and measurement noise are uncorrelated with each other. The case where the disturbance does not have a mean of zero can also be dealt with in this framework, provided that the mean is known. The same is true for the case of known input, and for the case where the noise processes are correlated in time, provided that a shaping filter for the noise is available. See Appendix A for details.

The optimization problem is defined as

$$\min_{\hat{x}_k} E\left(\|\tilde{x}_k\|^2\right) \;\;\equiv\;\; E\left(\tilde{x}_k' \tilde{x}_k\right), \quad k \in [1, K] \tag{2.18}$$

$$\text{subject to} \qquad \text{Eqs.(2.13)} - (2.17) \tag{2.19}$$

with

$$\tilde{x}_k \equiv x_k - \hat{x}_k \tag{2.20}$$

where $\hat{x}_k$ is a one-step predicted estimate, (i.e., a linear function of the past observations $y_{k-i}, 0 < i \le k$). Knowledge of the model equations (2.13)-(2.17) completely determines the correlation functions, though the converse of this statement does not hold as many models can give rise to the same correlation function. Thus, with the state-space description, the issue of model uncertainties is still present.

For the above problem, the one-step predicted estimate is given recursively by (See Section A.2 for the derivation).

$$\hat{x}_{k+1} \;=\; \tilde{A}_k \hat{x}_k + K_k y_k \tag{2.21}$$

$$\tilde{A}_k \;\equiv\; A_k - K_k C_k \tag{2.22}$$

where K_k is the *closed loop* Kalman gain given by

$$K_k = \left(A_k P_k C_k' + B_k D_k'\right)\left(C_k P_k C_k' + D_k D_k'\right)^{-1} \tag{2.23}$$

We assume that $(C_k P_k C_k' + D_k D_k')$ is invertible. This is true, for instance, when all measurements are noisy, so that $D_k D_k'$ is invertible, or when the term $C_k P_k C_k'$ is invertible. Here P_k is the *unconditional* one-step prediction error covariance

$$P_k \equiv E\left(\tilde{x}_k \tilde{x}_k'\right)$$

It is easy to show that the error vector $\tilde{x}_k$ satisfies the dynamics given by

$$
\begin{aligned}
\tilde{x}_{k+1} &= (A_k - K_k C_k)\tilde{x}_k + (B_k - K_k D_k) r_k \\
&= \tilde{A}_k \tilde{x}_k + \tilde{B}_k r_k
\end{aligned}
\tag{2.24}
$$

with

$$\tilde{B}_k \equiv B_k - K_k D_k \tag{2.25}$$

and $\tilde{A}_k$ as given by Eq.(2.22). The error covariance matrix P_k then satisfies the Riccati equation

$$
\begin{aligned}
P_{k+1} &= \tilde{A}_k P_k \tilde{A}_k' + \tilde{B}_k \tilde{B}_k' \quad k = 0, ..., N-1 \tag{2.26} \\
P_0 & \quad \text{given}
\end{aligned}
$$

The term closed loop gain is used to designate the matrix K_k because of its presence in the above error dynamics and covariance equations. It is clear that if P_0 is positive semi-definite, then so are the P_k's.

The one-step predicted estimate $\hat{x}_k$ is a function of the observations up to time $k-1$. The filtered estimate, $\hat{x}_{k|k}$, which is a function of the observations up to time k, or $y_0, ..., y_k$, is given by the measurement update equation in terms of the predicted estimate (Section A.3)

$$\hat{x}_{k|k} = \hat{x}_k + L_k \varrho_k \tag{2.27}$$

where L_k is the *open loop* Kalman gain, or simply the Kalman gain, and is given by

$$L_k = P_k C_k' \left(C_k P_k C_k' + D_k D_k'\right)^{-1} \tag{2.28}$$

and ϱ_k is the *innovation* (Section A.5)

$$\varrho_k \equiv y_k - C_k \hat{x}_k \tag{2.29}$$

The filtered estimate error covariance, $P_{k|k}$, is given by

$$P_{k|k} = P_k - P_k C_k' \left(C_k P_k C_k' + D_k D_k'\right)^{-1} C_k P_k \tag{2.30}$$

The Kalman filter and its error dynamics for the discrete-time system of Eqs.((2.13)-(2.17)) are given by Eqs.(2.21-2.30).

When the linear system of Eqs.(2.13)-(2.17) is a time-invariant one, then the subscript k can be dropped from the matrices, so that $A_k \equiv A$, $B_k \equiv B$, $C_k \equiv C$, and $D_k \equiv D$. The question then arises as to whether the Kalman filter is also time-invariant, meaning: are the gain and covariance matrices constant? The answer is that the solution of the Riccati equation (2.26) converges to a steady state, or $P_k \rightarrow P$, whenever the pair (A, B) is reachable and the pair (A, C) is observable. In that case, $K_k \rightarrow K$ and $P_{k|k} \rightarrow P_u$ as well. In addition, the error dynamics of Eq.(2.24) are stable even if the plant itself is not stable. Specifically, $\tilde{A}$ has all its eigenvalues inside the unit circle even if A has some outside. Notice that the reachability and observability conditions are sufficient, but not necessary, for the convergence of the Kalman filter matrices and the stability of the error dynamics[2] (Section A.6 and the references therein). Finally, at steady state, the Kalman filter

$$\hat{x}_{k+1} = (A - KC)\hat{x}_k + Ky_k$$

is the time domain equivalent of the matrix form of the Wiener filter of Eq.(2.8),

$$\mathcal{H}_W(z) = \left\{ S_{xy}(z) \left(S_{yy}^-(z) \right)^{-1} \right\}^+ \left(S_{yy}^+(z) \right)^{-1} \qquad (2.31)$$

In the frequency domain, the steady-state Kalman filter is also the H_2 optimal estimator in the sense that it minimizes the H_2 norm, or simply the 2-norm, of the transfer function $G_{\tilde{x}r}(z)$ between the input disturbance r and the error $\tilde{x}$. The 2-norm of a single-input single-output or a scalar transfer function $g(z)$ is given by

$$\|g\|_2 \equiv \left(\int_{-\pi}^{+\pi} g(e^{j\omega}) g(e^{-j\omega}) d\omega \right)^{\frac{1}{2}}$$

For a multi-input multi-output transfer function G, we have

$$\|G(e^{j\omega})\|_2 \equiv \left(\int_{-\pi}^{+\pi} \text{trace}\left(G(e^{j\omega}) G'(e^{-j\omega}) \right) d\omega \right)^{\frac{1}{2}} \qquad (2.32)$$

---

[2]The question of stability of the error dynamics for time-varying systems can also be addressed using more general definitions of reachability and observability, together with Lyapunov stability methods.

The Hardy space H_2 is the vector space of all 2-norm bounded transfer functions that are analytic outside the unit disc. The transfer function $G_{\tilde{x}r}$, which is stable if (A, B) is reachable and (A, C) is observable, belongs to H_2.

Now, if $S_{\tilde{x}\tilde{x}}(z)$ is the power spectral density of $\tilde{x}$, and $G_{\tilde{x}r}(e^{j\omega})$ has singular values[3] $\sigma_i(G), i = 1, ..., N$, then, it follows that when r is unit intensity white noise,

$$
\begin{aligned}
\lim_{k \to \infty} E\left(\tilde{x}'_k \tilde{x}_k\right) &= R_{\tilde{x}\tilde{x}}(0) \\
&= \frac{1}{2\pi} \int_{-\pi}^{+\pi} \text{trace}\left(S_{\tilde{x}\tilde{x}}(e^{j\omega})\right) d\omega \\
&= \frac{1}{2\pi} \int_{-\pi}^{+\pi} \text{trace}\left(G_{\tilde{x}r}(e^{j\omega})G'_{\tilde{x}r}(e^{-j\omega})\right) d\omega \\
&= \frac{1}{2\pi} \int_{-\pi}^{+\pi} \left(\sum_{i=1}^{N} \sigma_i^2\left(G_{\tilde{x}r}(e^{j\omega})\right)\right) d\omega \qquad (2.33) \\
&= \|G_{\tilde{x}r}\|_2^2 \qquad (2.34)
\end{aligned}
$$

In fact, the Kalman filter minimizes each component in the sum of Eq.(2.33). The H_2-norm therefore represents the energy in the error signal, and the Kalman filter can be seen as minimizing that energy. For this reason, the steady-state Kalman filter is called the H_2 optimal filter.

In this discussion, we assumed thus far that only the first two moments of the disturbance $r_k, k = 0, 1, ...,$ and the initial error $x - \hat{x}_0$, are known. We did not specify a density function for either one. If in addition we assume that the noise process and the initial error are both Gaussian, then the linearity of the model implies that at each time step k the state of the system, the observation, the one-step predicted and filtered estimates, their corresponding estimation error, and the innovation are all Gaussian. Moreover, the stochastic process x_k, described by Eq.(2.13), is Markov, meaning that $p(x_{k+1}|x_k, ..., x_0) = p(x_{k+1}|x_k)$. We therefore have a *Gauss Markov* process.

As a result, the Kalman filter acquires additional properties. First, its estimate computes the *conditional* mean of the state of the plant, or

$$
\hat{x}_k = E\left(x_k \mid y_0, \ldots, y_{k-1}\right) \qquad (2.35)
$$

---

[3]The singular values of a complex matrix A are the positive square roots of the eigenvalues of AA* or A*A, where A* is the complex conjugate of A. The singular value of a matrix transfer function is defined at each frequency.

$$\hat{x}_{k|k} \;=\; E\left(x_k \mid y_0, \ldots, y_k\right) \tag{2.36}$$

Since the mean of a random variable is also its minimum error variance estimate, it follows that, when the noise is Gaussian, the Kalman filter is among linear and nonlinear estimators the overall optimal filter in the least squares or minimum error variance sense. Another property follows from the fact that the mean of the Gaussian density function is also its mode, or most likely value. As a result, the Kalman filter gives the maximum a posteriori probability (MAP) estimate of x_k, conditioned on the available observations. Specifically,

$$\hat{x}_k \;=\; \arg\max_{x_k} p(x_k|y_0, \ldots, y_{k-1}) \tag{2.37}$$

$$\hat{x}_{k|k} \;=\; \arg\max_{x_k} p(x_k|y_0, \ldots, y_k) \tag{2.38}$$

Moreover, since the first two moments of a Gaussian uniquely determine the entire density function, it follows that the Kalman filter equations propagate the entire density function of the state of the system (Section A.4).

2.2.2   Beyond Linear Least Squares Estimation

Soon after its appearance, the Kalman filter was successfully used in many space applications, including the Apollo project, by groups at Draper Laboratory [14], [15], and NASA Ames. Extensions and improvements were also developed, such as the Kalman smoother [34], which solves the smoothing problem, and the square-root filter of Potter [3], which provides numerical stability by propagating the square-root of the covariance, instead of the covariance itself, as is done in the Riccati equation (2.26).

All of these algorithms are based on the same framework as the Kalman filter: 1) a linear plant and observation model; 2) least squares error or minimum error variance performance index; and 3) accurate knowledge of the model and disturbance statistics. Efforts to remove these restrictions have taken many directions. Among them:

1. *Nonlinear filtering*: The plant can be nonlinear, and the noise non-Gaussian. However, the plant model and noise statistics are assumed known. There are no optimal nonlinear filters that are numericaly implementable. In Section A.9, we extend the use of the Kalman filter as a suboptimal state estimator for nonlinear

systems. The resulting algorithm is the extended Kalman filter (EKF).

2. *Identification and adaptive filtering*: The plant is linear and the noise Gaussian, but some of the plant and/or noise parameters are uncertain. The goal is to estimate the unknown parameters along with the state of the system. The filter slowly adapts to the accurate model. The EKF can be used as a parameter estimator as well.

3. *Robust filtering*: The plant is linear, with bounded perturbations, and the noise statistics are unknown but bounded. Here the goal is to design filters with robustness to

   i) noise model uncertainties, or

   ii) noise and plant model uncertainties.

4. *Risk sensitive filtering*: The plant is linear and the noise is Gaussian, but the sum of squared-error performance index is replaced with the exponential of the sum of squared-error, which leads to filters that are conservative with respect to large error.

The estimation problem we are concerned with here falls within the third and fourth categories listed above. For the third category, we deal with norm-bounded perturbations and model uncertainties, as will be specified shortly. The plant is linear as in Eq.(2.13). The robust estimators we develop are applicable to a wide class of model uncertainties, including unmodeled dynamics and parametric uncertainties in the A, B, C, and D matrices of our plant. Regarding the fourth category, our objective is to establish the relationship between risk sensitive estimators and our class of robust estimators. Since the risk sensitive estimation problem is formulated in a stochastic setting, with the noise assumed Gaussian, this relationship can be seen as a stochastic interpretation of robust filtering.

Before we describe robust and risk sensitive estimation, we first motivate the problem by demonstrating the effect of modeling uncertainty on the performance of the Kalman filter.

2.2.3   Kalman Filter and Model Uncertainties

Assume that the true plant matrices in Eq.(2.13) are $A + \Delta A$, $B + \Delta B$, $C + \Delta C$, and $D + \Delta D$, but that the Kalman filter was designed with the assumption that the perturbations are absent. Then the actual state and error dynamics are given by

$$\begin{bmatrix} x_{k+1} \\ \tilde{x}_{k+1} \end{bmatrix} = \begin{bmatrix} A + \Delta A & 0 \\ \Delta A - K \Delta C & A - KC \end{bmatrix} \begin{bmatrix} x_k \\ \tilde{x}_k \end{bmatrix}$$
$$+ \begin{bmatrix} B + \Delta B \\ B - K(D + \Delta D) \end{bmatrix} r_k \qquad (2.39)$$

If $\Delta A = \Delta C \equiv 0$, and the system is observable and reachable, then the Kalman filter's error dynamics are stable. If $\Delta A \neq 0$, and the dynamic system is stable, observable and reachable, then the error still decays, even though the Kalman filter design was based on the wrong model.

If, however, the dynamic system is unstable, so that $x_k \to \infty$ with k, then any small but nonzero ΔA (and in reality there will always be such a ΔA), would give rise to growing error.

The above discussion indicates that if the dynamic system is unstable, the Kalman filter will always be unstable, no matter how small the nonzero perturbation ΔA is. If the dynamic system is stable, then the Kalman filter is not only stable, but robustly stable as well, at least for parametric uncertainties. *Kalman himself considered this robust stability property of his filter to be at least as important a contribution as the filter itself* [65].

But robust stability does not imply robust performance. Even if the Kalman filter is stable, its performance degrades considerably in the presence of modeling uncertainties. The following example illustrates this point.

Example:

Consider a nominal linear plant with matrices

$$A = \begin{bmatrix} 0.905 & 0 \\ 0 & 0.812 \end{bmatrix}, \quad B = \begin{bmatrix} 0.301 & 0 & 0 & 0 \\ 0 & 0.287 & 0 & 0 \end{bmatrix},$$

$$C = \begin{bmatrix} 1 & 0 \\ 0 & 1 \end{bmatrix}, \quad D = \begin{bmatrix} 0 & 0 & 0.1 & 0 \\ 0 & 0 & 0 & 0.1 \end{bmatrix}$$

and a perturbed plant with matrices

$$A = \begin{bmatrix} 0.827 & 0 \\ 0 & 0.980 \end{bmatrix}, \quad B = \begin{bmatrix} 0.231 & 0 & 0 & 0 \\ 0 & 0.454 & 0 & 0 \end{bmatrix}$$

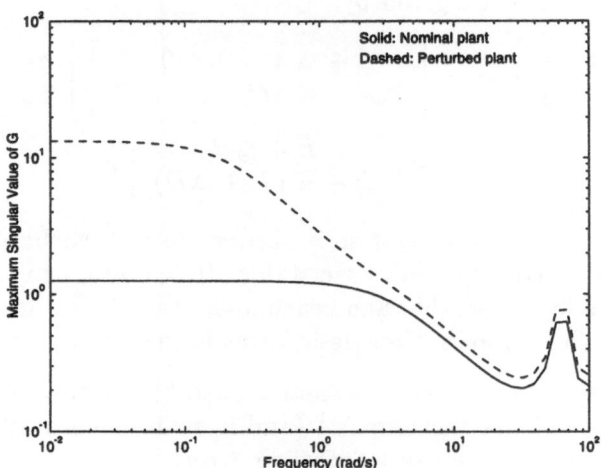

Figure 2.1: Kalman filter's frequency domain performance for nominal and perturbed system.

and the same C and D matrices. A Kalman filter, designed based on the nominal plant, is used for both plants. Figure 2.1 shows the maximum singular value of the transfer function from input noise to state estimation error for both plants. With the nominal plant, the Kalman filter minimizes the area under the singular value curve (See Eqs.(2.33-2.34)). Thus, the solid curve represents what the performance is believed to be, while the dashed curve represents what the performance actually is.

Figure 2.2 shows the effect of model uncertainties from a stochastic point of view. The figure shows the probability density function of the cost function $J = \sum_{k=0}^{N} (x_k - \hat{x}_k)' (x_k - \hat{x}_k)$ for the Kalman filter with both the nominal and perturbed plant dynamics. Appendix B explains how the error density functions of Figure 2.2 are computed. These plots illustrate the large increase in the probability of a large error for the Kalman filter in the presence of plant uncertainty.   ◇

Overall, this discussion shows that while the Kalman filter is perfectly tuned to a nominal plant, it behaves poorly for a perturbed one. It may be preferable to sacrifice some nominally optimal performance in order to mitigate the effect of modeling error. This is precisely the aim of robust estimation!

This example will be used in Chapter 3 and again in Chapter 4 to

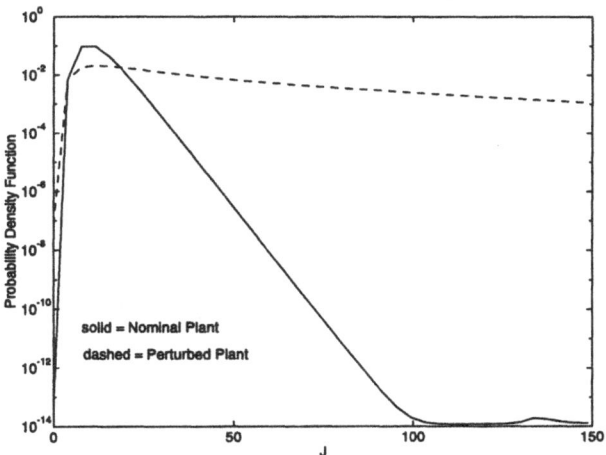

Figure 2.2: Error probability density function comparison for Kalman filter.

illustrate the effectiveness of robust filters with modeling uncertainties.

2.2.4   Robust Estimation and the Small Gain Theorem

As mentioned before, the goal in robust control and estimation is to design compensators and filters that take into account model uncertainties. For the filtering problem, Figure 2.3 shows a plant P whose inputs are the initial condition x_0, and the signal r, which represents the disturbance and the known inputs. The initial error estimate is $x_0 - \hat{x}_0$. The plant's output is y. The block Δ represents the model uncertainties, while η and ϵ are the interaction between the plant and the uncertainty. The plant's output is fed into the filter F, along with the initial estimate $\hat{x}_0$. The filter's output is the estimate $\hat{x}$ and the estimation error e. The estimation error is given by $e = M(x - \hat{x})$, where M is some weighting matrix.

Our goal is to design a filter that bounds the error norm for an entire class of perturbations. Before specifying this objective more precisely, we provide some mathematical preliminaries.

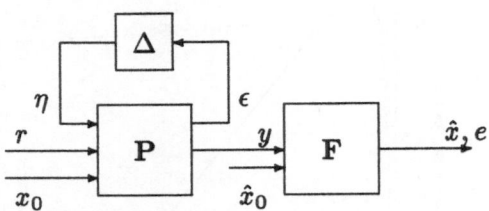

Figure 2.3: The robust estimation problem

Define the p norm of a vector $\zeta = (\zeta(1), ..., \zeta(n)) \in R^n$ as

$$\|\zeta\|_p \equiv \left(\sum_{j=1}^{n} |\zeta(j)|^p \right)^{1/p} \tag{2.40}$$

for $1 \leq p < \infty$. The infinity norm of ζ is defined as

$$\|\zeta\|_\infty \equiv \max_{i \in [1,n]} \zeta(i) \tag{2.41}$$

For a sequence $x = x_1, x_2, ...$, with $x_i \in R^n, i = 1, 2, ...$, we define the p and infinity norms of the sequence as, respectively,

$$\|x\|_p \equiv \left(\sum_{i=1}^{\infty} \|x_i\|_p^p \right)^{1/p} \tag{2.42}$$

$$\|x\|_\infty \equiv \sup_{i} \|x_i\|_\infty \tag{2.43}$$

The space ℓ_p^n, or simply ℓ_p, for $1 \leq p < \infty$, is defined as

$$\begin{aligned} \ell_p &\equiv \left\{ x = (x_1, x_2, ...) \mid x_i \in R^n, i = 1, 2, ..., \text{ and } \|x\|_p < \infty \right\} \\ &\equiv \ell_p^n \end{aligned} \tag{2.44}$$

For $p = \infty$, the space ℓ_∞ is defined as

$$\begin{aligned} \ell_\infty &\equiv \left\{ x = (x_1, x_2, ...) \mid x_i \in R^n, i = 1, 2, ..., \text{ and } \|x\|_\infty < \infty \right\} \\ &\equiv \ell_\infty^n \end{aligned} \tag{2.45}$$

To specify the error bound of an estimator, we rely on the concept of an operator or **induced norm**. The induced norm of a mapping is a measure of the maximum amplification the mapping can exert on a bounded input signal. Let $\mathcal{G}$ be a mapping between two normed spaces X and Y, with respective norms $\|.\|_\alpha$ and $\|.\|_\beta$. Then the induced (β, α) norm of $\mathcal{G}$ is given by

$$\|\mathcal{G}\|_{i(\beta,\alpha)} \equiv \sup_{r \neq 0} \frac{\|\mathcal{G}r\|_\beta}{\|r\|_\alpha} \tag{2.46}$$

$$= \sup_{\|r\|_\alpha \leq 1} \|\mathcal{G}r\|_\beta \tag{2.47}$$

$$= \sup_{\|r\|_\alpha = 1} \|\mathcal{G}r\|_\beta \tag{2.48}$$

It is easy to verify the equivalence of the above definitions. If $\alpha = \beta$, we simply write $\|\mathcal{G}\|_{i\alpha}$. Often, however, the entire subscript is omitted from the notation when a statement is applicable to any norm, or when it is obvious from the context of the discussion which norm is implied.

Induced norms are of interest because they represent a worst-case objective or gain, as indicated by the presence of the supremum in the above definitions. Bounding or minimizing an induced norm, therefore, guarantees worst-case results. In contrast, the H_2 norm, defined in Eq.(2.32), is not an induced norm, as it measures the total energy of the mapping.

For the robust estimation problem of Figure 2.3, the objective is to find a filter F such that the induced norm of the mapping from the inputs r, x_0, and $x_0 - \hat{x}_0$, to the estimation error, e, is bounded for all possible perturbations Δ. It is assumed that the inputs and perturbations are norm bounded. If the initial condition and error are included in the definition of the signal r, then we can formulate the following performance criterion

$$\|\mathcal{G}\| \equiv \sup_{\|r\|=1} \|e\| < 1$$

$$\forall \Delta \ \ni \ \|\Delta\| \equiv \sup_{\|\epsilon\| \neq 0} \frac{\|\eta\|}{\|\epsilon\|} \leq 1 \tag{2.49}$$

Note that it is possible to rescale or normalize the perturbation in order to obtain a bound of one on its induced norm.

If $r, e \in \ell_2$, then the induced 2-norm is given by

$$\|\mathcal{G}\|_{i2} \equiv \sup_{r \neq 0} \frac{\|\mathcal{G}r\|_2}{\|r\|_2} \tag{2.50}$$

In the frequency domain, if the mapping $\mathcal{G}$ is identified with a stable discrete-time transfer function, then it can be seen as an element of the Hardy space H_∞, which is the space of transfer functions who are analytic outside the unit disc, and whose induced 2-norm is bounded. In the frequency domain, the induced 2-norm is identified with the H_∞ norm. This norm is a function of the maximum singular value of the same transfer function. Specifically, if G is the transfer function corresponding to the map $\mathcal{G}$, then (see [11], [20] for the continuous-time case and Section 3.4 for the discrete-time case),

$$
\begin{aligned}
\|\mathcal{G}\|_{i2} &= \|\mathcal{G}\|_\infty \\
&\equiv \sup_\omega \sigma_{\max}\left(G(e^{j\omega})\right)
\end{aligned}
\tag{2.51}
$$

The induced infinity norm is another induced norm that can be used for the design of compensators and filters [19]. If $r \in \ell_\infty$, this norm is defined as

$$
\|\mathcal{G}\|_{i\infty} = \sup_{r \neq 0} \frac{\|\mathcal{G}r\|_\infty}{\|r\|_\infty}
\tag{2.52}
$$

Filters based on this norm for linear time-invariant systems at steady state have been derived in [109], but they are robust to noise model uncertainties only. In this book, we will be concerned with the 2-norm, as the resulting filters are generalizations of the Kalman filter, and they can be time-varying as well as time-invariant. Time-varying filters are useful in failure detection, since failures are a transient phenomenon. Thus, in Eq.(2.49), we assume that all the signals are in ℓ_2, the linear space of square summable sequences. The ℓ_2 and induced-2 norm are therefore implied.

In the presence of model uncertainty, induced norms are also of interest because of the **small gain theorem** ([20],[75]). This theorem states that a system with a feedback loop composed of stable mappings is itself stable provided the product of all the mappings' induced norms is less than unity. More simply, the theorem provides a sufficient condition for the stability of the feedback system representation of Figure 2.4. Most feedback systems can be represented in the form of that figure. The small gain theorem is important in control because it guarantees robust stability, or stability in the presence of uncertainties. It can also be used, as will be seen below, to guarantee robust performance, which is important to estimation. In Section 2.2.3, we mentioned that if a plant and estimator dynamics are stable, then the estimation error dynamics are robustly stable provided the estimator is stable. For estimators, therefore, robust stability is not an issue.

Robust performance, however, is. The estimator's robust performance criterion, given by Eq.(2.49), is the worst case gain, over the entire range of plant model uncertainty, of the closed-loop mapping whose input is the initial estimation error, the initial condition, and the disturbance, and whose output is the estimation error.

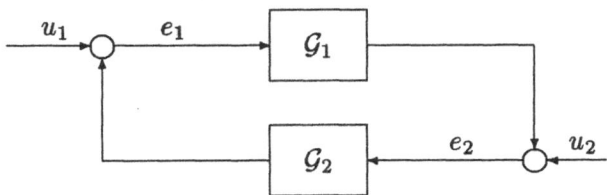

Figure 2.4: Feedback configuration for the small gain theorem

Theorem 2.1 *Let $\mathcal{G}_1 : \ell_p^n \rightarrow \ell_p^m$ and $\mathcal{G}_2 : \ell_p^m \rightarrow \ell_p^n$ be two stable operators with bounded gains. Then, the closed loop system is stable if $\|\mathcal{G}_1\|\|\mathcal{G}_2\| < 1$.*

Proof: We assume that for any pair of inputs u_1 and u_2 (Figure 2.4), there exists a unique pair of outputs e_1 and e_2. This assumption holds whenever the mapping $\mathcal{G}_1$ and $\mathcal{G}_2$ are linear. Noting that $e_1 = \mathcal{G}_2 e_2 + u_1$ and $e_2 = \mathcal{G}_1 e_1 + u2$, the norms of the ouptuts are bounded by (again, we omit subscripts from the notation for norm)

$$\|e_1\| \leq \|\mathcal{G}_2\|\|e_2\| + \|u_1\|$$
$$\|e_2\| \leq \|\mathcal{G}_1\|\|e_1\| + \|u_2\|$$

Substituting the second equation into the first, we have

$$\|e_1\| \leq \|\mathcal{G}_2\| (\|\mathcal{G}_1\|\|e_1\| + \|u_2\|) + \|u_1\|$$

Since $\|\mathcal{G}_1\|\|\mathcal{G}_2\| < 1$, it follows that

$$\|e_1\| \leq \left(\frac{1}{1 - \|\mathcal{G}_1\|\|\mathcal{G}_2\|}\right) (\|\mathcal{G}_2\|\|u_2\| + \|u_1\|)$$

The inequality

$$\|e_2\| \leq \left(\frac{1}{1 - \|\mathcal{G}_1\|\|\mathcal{G}_2\|}\right) (\|\mathcal{G}_1\|\|u_1\| + \|u_2\|)$$

follows from a similar argument.   ◇

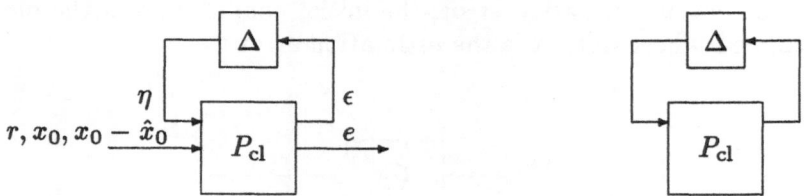

Figure 2.5: Robust performance (left) and robust stability (right).

2.2.5   Robust Stability and Robust Performance for Estimation

We now discuss how the small gain theorem is used for *robust stability* and *robust performance*. If the plant and filter form one augmented system, then the system on the left-hand-side of Figure 2.5 is the closed loop form of the system of Figure 2.3. Robust stability is the requirement that the output e settles to zero if all inputs are set to zero. Thus, robust stability requires that the system on the right-hand-side of Figure 2.5 be stable. If the uncertainty Δ in that figure is normalized to have an induced norm of unity, then the small gain theorem requires that the induced norm of the mapping between the input and the output to P_{cl} in that figure be less than unity for robust stability to hold. As mentioned before, if the plant and filter are stable, then, and only then, is P_{cl} stable, so that robust stability is not issue in filter design.

The motivation behind the design of robust filters is robust performance. We refer again to the system on the left-hand-side of Figure 2.5. For robust performance, we require conditions on P_{cl} such that the system is stable, and, in addition, that the mapping $G_{e(r,x_0,x_0-\hat{x}_0)}$, have a bounded induced norm

$$\|G_{e(r,x_0,x_0-\hat{x}_0)}\| < 1 \qquad (2.53)$$

To obtain conditions for the robust performance problem, we transform it into a robust stability problem by adding a fictitious perturbation Δ_f, as is done in Figure 2.6. A sufficient condition for robust performance is the robust stability of the system in Figure 2.6. If $\|\Delta\| = \|\Delta_f\| = 1$,

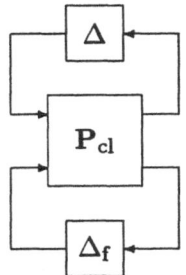

Figure 2.6: Robust stability for robust performance

then robust stability is achievable if the induced norm of the mapping between the augmented input and the augmented output of P_{cl} is less than one. The concept of *robust stability for robust performance* is the key to applying the small gain theorem to the design of robust filters.

In Chapter 3 and Appendix C, we use this concept to obtain sufficient conditions for the solution of the robust estimation problem of Eq.(2.49). To obtain a filter that satisfies the sufficient conditions, we formulate a game problem, where one player is the state estimate, and the adversary is the disturbance, initial condition, initial error, and plant perturbation. The game problem, in turn, requires the solution of two Riccati equations. If there is no model uncertainty, then only one Riccati equation needs to be solved. The result is a family of estimators of which the Kalman filter is but one member.

2.2.6   Further Discussion on Robust Estimation and Control

As mentioned earlier, we will be concerned with linear systems and the induced 2-norm. For linear time-invariant systems, the mapping $\mathcal{G}$ can be seen as a transfer function, and the induced 2-norm is the H_∞ norm of that transfer function. In 1981, Zames [125] proposed minimizing this norm. Zames' approach relied on an input-output operator theoretic setting [31] which was computationally difficult.

By 1984, Glover [38] made the first attempts at developing state-space solutions to the H_∞ control problems. Although his state-space solution was a vast improvement over the operator theory based ones, the approach was still computationally difficult and required higher or-

der compensators. Later, in 1989, Doyle, Glover et al. [24] developed a
state-space solution that required solving two Riccati equations. This
approach proved attractive from a computational point of view. Con-
tributions to the state-space solutions are also found in [39, 75, 106].

Robust control and estimation algorithms based on the small gain
theorem guarantee stability and performance for a linear system whose
uncertainties are defined by an induced norm bound. Characterizing
the uncertainties by their norm bound alone is, however, too conservative, since such a characterization leads to a larger than desired class of
uncertainties. For instance, uncertainties in Figure 2.6 have the block
diagonal form

$$\Delta_a = \begin{bmatrix} \Delta_f & \\ & \Delta \end{bmatrix} \tag{2.54}$$

The small gain theorem does not take this form into account, and provides sufficient conditions for any uncertainty block. As such, it is
conservative. A less conservative measure of robust stability and performance is the structured norm, which takes into account the block
diagonal structure of the uncertainties. In 1982, Doyle [25] introduced the structured singular value, μ, for linear time-invariant systems at steady-state (see also [26]). A robust H_∞/μ estimator is derived in [4], [5] for continuous-time linear time-invariant systems at
steady-state. Note that even μ is conservative when robust stability is
used for robust performance, since we are assuming a fictitious perturbation. Note also that when the only model perturbations are due to
real parameter errors, then defining uncertainties by their structured
norm is still conservative, since complex perturbations are included.
Reducing the conservatism of robust controllers and estimators by taking the structure of the uncertainty into account is the subject of much
research (e.g., [53], [123]). A survey of robust control methods is given
in [23], and a diversified collection of robust control problems is given
in [23], [96].

In this book, we rely on the induced-2 or H_∞ norm to derive filters
that are robust to a general class of model uncertainties for discrete and
continuous-time linear systems. The systems can be time-varying or
time-invariant over a finite or infinite-time horizon, with an arbitrary
initial condition.

Estimators or controllers with robustness to model uncertainties
that are not described by induced norm bounds also exist. An example
is the work of Schick [92], where estimators with robustness to outliers
in the observation noise are developed. The work in [92] is based on the

theory of robust statistics developed by Huber [54]. Specifically, it is
assumed that the observation noise (the term $v_k = D_k r_k$) in Eq.(2.14)
is white, independent of the process noise ($D_k B_k' = 0$), and has a
contaminated or heavy-tailed Gaussian density of the form

$$(1 - \epsilon)\mathcal{N}(0, \Sigma) + \epsilon g \qquad (2.55)$$

where ϵ is a very small constant, $\mathcal{N}(0, \Sigma)$ denotes the Gaussian density
function with mean 0 and covariance Σ, and g is an arbitrary density
function with a *very large* mean. As a result, the least squares estimate
produced by the Kalman filter gives rise to an unacceptably large error.
Schick derives an approximation of the density function of the state
conditioned on the observations, which he uses to obtain a recursive
filter for an approximate maximum a posteriori estimate of the state.

Other examples of representations of model uncertainty are also
found in the control literature, such as the work in [81], which is concerned with robustness towards parametric uncertainties only. Here
the parameter perturbations are constrained to be real. The seminal
work for this research is Kharitonov's result on interval polynomials
[67], which was followed by results on a variety of robust stability tests
[8] (See also [94] for a tutorial on this approach). These results are,
however, of limited practical value, as the class of problems studied was
restrictive, and the computational burden prohibitive.

Other research in robust control is that of [101], which relies on a
stochastic characterization of plant model uncertainties.

2.2.7   Risk Sensitive Control and Estimation

In risk sensitive estimation, the sum of squared error criterion of
Eq.(2.18) is replaced with the following one

$$\mathcal{L} \equiv \theta^{-1} \log \, E\left[\exp\left(\theta J\right)\right] \qquad (2.56)$$

where θ is a positive constant, and

$$J \equiv \sum_{k=1}^{N} e_k' e_k \qquad (2.57)$$

$$e_k \equiv M_k(x_k - \hat{x}_k)$$

subject to the same constraints (2.13-2.17). Furthermore, the noise is
assumed Gaussian. The cost function J can also be defined in terms
of the filtered estimate $\hat{x}_{k|k}$.

Here, we assume that the noise is *Gaussian*. As a result of the exponentiation, higher moments are included in the cost. To see this, consider the expansion of $\mathcal{L}$ around $\theta = 0$ (See [112], or Chapter 4)

$$\mathcal{L} = E(J) + \frac{\theta}{2}\text{var}(J) + O(\theta^2)$$

Since J is a function of the sum of squared error, the second term is on the order of the fourth moment of the error. This concern for large error gives a clue as to the relationship between risk sensitive and robust estimation, when the latter is based on the induced 2-norm, or H_∞, optimization. In Eq.(2.50), where the induced 2-norm is defined, we are concerned with the error's largest possible 2-norm. In Chapter 4, the relationship between these two classes of problems is demonstrated using a subtle relationship between expectation and maximization, both of which are taken with respect to the noise signal r. We also note that Eq.(2.56) represents a class of problems parametrized by θ. For $\theta = 0$, we have $\mathcal{L} = E(J)$, and we recover the minimum variance estimation problem, the solution of which is the Kalman filter.

Risk sensitive control was first considered in 1973 by Jacobson [57], who treated the case of perfect state observations. In fact, Jacobson pointed out the relationship between exponential Gaussian objectives and deterministic differential games (of which H_∞ is a special case, as will be seen in the next chapter). Subsequently, in 1974, Speyer, Deyst, and Jacobson [100] derived results for special cases of the general risk sensitive control problem with imperfect observations. Kumar and Van Schuppen obtained related results in 1981 [70].

Though the authors in [57] and [100] did not obtain a complete solution to the risk sensitive control problem, their effort was nevertheless a pioneering one, as they have considered a new objective function that is more general than the error variance. Whittle obtained the complete solution in 1981 [111]. In his analysis, a risk sensitive certainty equivalence principle is derived, from which a differential game interpretation follows.

The work mentioned thus far dealt with the discrete-time case. The continuous time case was treated by Bensoussan and Van Schuppen [17], who found results analogous to those of [71]. The complete solution for the continuous-time case, as well as the entire subject of risk sensitive control for nominal plants, is treated in the book by Whittle [112]. More recently, Speyer, Fan, and Banavar [99] considered the estimation problem.

In Chapter 4, we define a risk sensitive estimation problem for uncertain plants, and we show how it is related to the robust game theoretic estimation problem of Chapter 3.

2.3   Failure Detection and Isolation

In signal detection problems, noisy observations are used in a hypothesis test in order to determine whether they contain a signal. In communications, for instance, the objective is to detect the existence of a message in the midst of noise, while in control systems, the signal of interest can be a failure. Whatever the application, detection algorithms require a pre-processing of the raw observations obtained. For control systems, this is often done using a linear filter. Before discussing the role of linear filters in failure detection and isolation problems, we first give a brief description of the problem.

A failure detection and isolation (FDI) algorithm has in general two objectives: 1) To detect the failure, and 2) To isolate the failed component. A possible additional objective is failure identification, or estimating the failure.

Failure detection depends on hardware or analytical redundancy to detect anomalies in the system's behavior. An example of hardware redundancy is the voting method, which involves replication of physical devices. For example, if each required sensor is triplicated, the outputs of like sensors can be compared directly, and if there is a discrepancy between the signal of any one of the three sensors and the other two, then a failure is both detected and isolated. Because of the increase in cost as well as the increased weight and space requirement resulting from hardware redundancy, these methods are not attractive for many applications. Moreover, it is generally not possible to directly measure the output of many components such as actuators. For these reasons, many systems rely on indirect analytic methods, where knowledge of the plant dynamics is used to check for inconsistencies in the plant's behavior in order to detect and isolate failures.

Analytic approaches to fault detection and isolation (FDI) have been the subject of intensive research. A detailed description of this research will not be given here as this is done in many surveys, such as those by Willsky, [115], Sermann [56], Basseville [12], Patton and Chen [86], Frank [35], and Gertler [37]. The last survey is of particular interest as it focusses on the role of linear filters as residual generators in FDI algorithms. The subject is also treated in the textbook by Basseville and Nikiforov [13].

A failure in a plant or dynamic system can be seen as a sudden change in one or more of the plant parameters, or input signals. Such changes can be classified into two categories [13]: Additive failures, and non-additive failures. *Additive failures* produce a change in the mean of the observations. These can correspond to faults in the actuators or sensors, or to unanticipated failures, i.e., failures which cannot be assigned to any component, such as a submarine getting caught in a net. An example of work on unanticipated failures is found in [30]. *Non-additive failures* result in changes in either the variance of the output, or its spectral content. Such failures can occur in either the actuators or sensors, or the internal structure of the plant. Methods for dealing with non-additive failures are found, for instance, in [13].

We will be mostly concerned with *additive failures*. Two early works on the subject are the Detection Filter of Beard [16] and Jones [59], and the Generalized Likelihood Ratio Test (GLRT) of Willsky and Jones [114]. In both of these, a filter processes the measurements and supplies a residual to a threshold selector. In Detection Filter design, the idea is to choose the filter gain so that the residual output has a different fixed direction for each candidate component failure. While the detection filter is geometric in nature, the GLRT is a statistical test that looks for a change in the statistical properties of the filter's output. As our robust FDI algorithm of Chapter 5 is a extension of the GLRT, we will focus attention on this test.

In Section 2.3.1, we formulate the detection and isolation test for additive failures, and describe the algorithm. In Section 2.3.2, we formulate the problem for non-additive failures, and briefly discuss it. The effect of model uncertainty on the GLRT is the subject of Section 2.3.3, and a preliminary discussion on robust failure detection and isolation, the focus of Chapter 5, appears in Section 2.3.4.

2.3.1   Kalman Filters in FDI Algorithms: The GLRT

Our objective in this section is to develop an algorithm for detecting and isolating additive failures in linear Gaussian dynamic plants, when the noise statistics and the plant model are known. The algorithm is to be made suitable for online implementation.

The no-failure (H_0) and failure (H_1) hypotheses are

$$H_0 \quad : \quad x_{k+1} = A_k x_k + B_k r_k + U_k u_k \qquad (2.58a)$$

$$y_k = C_k x_k + D_k r_k + W_k u_k \qquad (2.58b)$$

$$H_1 \quad : \quad x_{k+1} = A_k x_k + B_k r_k + U_k u_k + F_k s_{k-\tau^*} \cdot \nu \qquad (2.58c)$$

$$y_k = C_k x_k + D_k r_k + W_k u_k + V_k s_{k-\tau^*} \cdot \nu \qquad (2.58d)$$

where x_k is the state of the system, y_k is the measurement, u_k represents the known inputs, and r_k is the vector of all exogenous disturbances: process and sensor noise. The random components of r_k are assumed to be *Gaussian* of zero mean and unit covariance. The matrices F_k and V_k describe the way the failure is injected into the system. For instance, if the failed component is an actuator, then F_k is the column vector of U_k representing that actuator in the plant dynamics. The parameter τ^* is the *unknown* time at which the failure occurs, the function s_k is the unit step function, and ν is the magnitude of the step failure, which is generally not known. As described in [42], [47], and [114], the GLRT considers only unimodal failures, such as a jump, a step, a ramp, etc. Our description of the test resembles more that of [47].

For observations $y_{k-L}, ..., y_k$ over a finite interval of length $L+1$ time steps, the likelihood ratio [88] of the above hypothesis test is given by

$$\Lambda_k(\tau^*, \nu) = \frac{p(y_{k-L}, ..., y_k \mid H_1, \tau^*, \nu)}{p(y_{k-L}, ..., y_k \mid H_0)} \qquad (2.59)$$

Because both the failure time, τ^*, and its magnitude, ν, are unknown, we replace the above likelihood ratio by the generalized likelihood ratio (GLR)

$$\Lambda_k^* = \max_{\tau^* \in [k_0 - L, k]} \max_{\nu} \Lambda_k(\tau^*, \nu) \qquad (2.60)$$

We would like to compare the GLR to a threshold at each time k. If it exceeds the threshold, then a failure is declared. The comparison can be made at each time k by running a sliding window $[k - L, k]$, $[k - L + 1, k + 1]$, etc. Threshold selection is a tradeoff between the *false alarm rate*, and *the detection time*. The lower the threshold, the higher the false alarm rate, but the quicker is the detection, and vice versa. The false alarm rate is a function of the window length as well, while the detection speed is also a function of the window length and the failure magnitude.

To implement the above test, we need to compute Λ_k^* at each step, and compare it to a threshold. Three difficulties accompany this implementation. First, the test requires two maximizations at each step. The maximization over ν is doable, as we shall see shortly, but the maximization over τ^* is computationally prohibitive, and requires a brute force approach. Specifically, we need to compute the ratio for each possible $\tau^* \in [k - L, k]$ at each time step k. Second, even if τ^* is fixed, the ratio cannot be computed recursively as the observations $y_{k-L}, \ldots, y_k$ are not statistically independent. Third, it is not possible to obtain an analytical expression for the threshold as a function of the false alarm rate and detection speed.

Threshold selection is usually achieved after extensive simulation experiments, where the window length and the threshold level are varied. We now deal with the two other difficulties.

Since the observations $y_{k-L}, \ldots, y_k$ are not independent, we would like to replace them by their corresponding innovations (Section A.5), $\varrho_{k-L}, \ldots, \varrho_k$, which are statistically independent of one another. Thus, we have

$$\varrho_k = y_k - C_k \hat{x}_k - W_k u_k \tag{2.61}$$

where $\hat{x}_k \equiv E(x_k \mid y_0, \ldots, y_{k-1})$. The hypotheses test of Eqs.(2.58a-2.58d) can then be expressed in terms of the innovation

$$H_0 \quad : \quad \varrho_k = \varrho_{ok} \tag{2.62a}$$
$$H_1 \quad : \quad \varrho_k = \varrho_{ok} + G_k(\tau^*)\nu \tag{2.62b}$$

for all $k \in [k - L, k]$. Here ϱ_{ok} is the innovation in the absence of failure, and $G_k(\tau^*)$ is the additive failure signature, which can be recursively computed, as shown below. To obtain the innovation, we design a Kalman filter for the no-failure model of Eqs.(2.58a-2.58b)

$$\hat{x}_{k+1} = A_k \hat{x}_k + K_k \varrho_k + U_k u_k \tag{2.63a}$$
$$K_k = A_k P_k C_k' (D_k D_k' + C_k P_k C_k')^{-1} \tag{2.63b}$$
$$P_{k+1} = (A_k - K_k C_k) P_k (A_k - K_k C_k)' $$
$$+ (B_k - K_k D_k)(B_k - K_k D_k)' \tag{2.63c}$$

For simplicity, we have assumed that the process and measurement noise are uncorrelated, i.e., $B_k D_k' = 0$. The above filter gives the a priori estimate at time k. In its whitening filter form, the Kalman filter can be written as

$$\hat{x}_{k+1} = (A_k - K_k C_k)\hat{x}_k + K_k y_k + U_k u_k \tag{2.64a}$$
$$\varrho_k = y_k - C_k \hat{x}_k - W_k u_k \tag{2.64b}$$

From the above equations, it can be seen that the observation can be reconstructed from the innovation, and vice-versa. Both of these sequences contain the same statistical information. The two hypothesis tests of Eqs.(2.58a-2.58d) and Eqs.(2.62a-2.62b) are therefore equivalent. The advantage of the latter is that, unlike the observations, the innovations are statistically independent, a fact that will be exploited shortly to make the GLRT recursive.

Now, the failure signature, $G_k(\tau^*)$, can be computed as the output of a linear system

$$\Upsilon_{k+1}(\tau^*) = (A_k - K_k C_k)\Upsilon_k(\tau^*) + [F_k - K_k V_k]s_{k-\tau^*} \quad (2.65)$$
$$G_k(\tau^*) = C_k \Upsilon_k(\tau^*) + V_k s_{k-\tau^*} \quad (2.66)$$

The likelihood ratio for our hypothesis test is

$$\Lambda_k(\tau^*, \nu) = \frac{p(\varrho_{k-L}, ..., \varrho_k \mid H_1, \tau^*, \nu)}{p(\varrho_{k-L}, ..., \varrho_k \mid H_0)}$$
$$= \prod_{j=k-L}^{j=k} \frac{p(\varrho_j \mid H_1, \tau^*, \nu)}{p(\varrho_j \mid H_0)} \quad (2.67)$$

where the last equality follows from the independence of the innovation process.

Taking the log of the above ratio, it follows from the Gaussian assumption that

$$\lambda_k(\tau^*, \nu) \equiv \log \Lambda_k(\tau^*, \nu)$$
$$= \nu\chi_k(\tau^*) - \frac{1}{2}\nu^2 S_k(\tau^*) \quad (2.68)$$

where

$$\chi_k(\tau^*) = \sum_{j=\tau^*}^{k} G'_j(\tau^*)R_j^{-1}\varrho_j \quad (2.69)$$
$$R_j = C_j P_j C'_j + D_j D'_j \quad (2.70)$$
$$S_k(\tau^*) = \sum_{j=\tau^*}^{k} G'_j(\tau^*)R_j^{-1}G_j(\tau^*) \quad (2.71)$$

The *generalized* log likelihood ratio is given by

$$\ell_k = \max_{\tau^* \in [k-L, k]} \max_{\nu \in R} \lambda_k(\tau^*, \nu) \quad (2.72)$$

It is clear that $\lambda_k(\tau^*, \nu)$ is additively recursive. We first maximize the above expression with respect to ν at each step k. Thus, the step failure magnitude ν is replaced by its maximum likelihood estimate

$$\hat{\nu}_k(\tau^*) = \frac{\chi_k(\tau^*)}{S_k(\tau^*)} \tag{2.73}$$

Finally, if we define

$$\overline{\lambda}_k(\tau^*) = \max_{\nu \in R} \lambda_k(\tau^*, \nu) \tag{2.74}$$

then, by substituting into Eq.(2.68), we have, apart from a constant multiple (that can be incorparated into the threshold)

$$
\begin{aligned}
\ell_k &= \max_{\tau^* \in [k-L,k]} \overline{\lambda}_k(\tau^*) \\
&= \max_{\tau^* \in [k-L,k]} \frac{\chi_k^2(\tau^*)}{S_k} \tag{2.75} \\
&= \max_{\tau^* \in [k-L,k]} S_k \hat{\nu}_k^2(\tau^*) \tag{2.76}
\end{aligned}
$$

As mentioned earlier, the maximization over the failure time τ^* is a computationally burdensome problem, as only a brute force approach is possible. Specifically, at each time k, we must compute $\overline{\lambda}_k(\tau^*)$ for every $\tau^* \in [k - L, k]$ in order to determine ℓ_k and compare it to a threshold. One way of simplifying the problem is to simply drop the failure time from the test hypothesis. For any window $[k - L, k]$ of data, the failure is assumed either present or absent for the entire time interval. This is equivalent to fixing $\tau^* \equiv k - L$ in the H_1 hypothesis. The generalized likelihood ratio of Eqs.(2.75,2.76) simplifies then to

$$
\begin{aligned}
\ell_k &= \frac{\chi_k^2}{S_k} \tag{2.77} \\
&= S_k \hat{\nu}_k^2 \tag{2.78}
\end{aligned}
$$

Note that S_k^{-1} is essentially the error variance of the estimate $\hat{\nu}_k$. The above equations suggest that there are two ways to compute the likelihood ratio. One way, suggested by the first equation (2.77) and already discussed, is to obtain the residual or innovation ϱ_k of a Kalman filter based on the no-failure hypothesis. This is the approach followed in [42], [47], and [114].

The second equation (2.78), however, suggests another procedure, based on a direct computation of the failure's estimate, $\hat{\nu}$. As such, one can view the failure detection test as a two-sided hypothesis test on ν, or

$$H_0 \ : \ \nu = 0 \tag{2.79a}$$
$$H_1 \ : \ \nu \neq 0 \tag{2.79b}$$

with the quantity of Eq.(2.78) as the chosen statistic for that test. The maximum likelihood estimate $\hat{\nu}$ is obtained using a transient Kalman filter for the augmented system

$$\begin{bmatrix} x_{k+1} \\ \nu_{k+1} \end{bmatrix} = \begin{bmatrix} A_k & F_k \\ 0 & 1 \end{bmatrix} \begin{bmatrix} x_k \\ \nu_k \end{bmatrix} + \begin{bmatrix} U_k \\ 0 \end{bmatrix} u_k + \begin{bmatrix} B_k \\ 0 \end{bmatrix} r_k \tag{2.80}$$

$$y_k = \begin{bmatrix} C_k & 0 \end{bmatrix} \begin{bmatrix} x_k \\ \nu_k \end{bmatrix} + D_k r_k \tag{2.81}$$

with initial mean and covariance

$$E\left(\begin{bmatrix} x_{k-L} \\ \nu_{k-L} \end{bmatrix} \right) = \begin{bmatrix} \hat{x}_{k-L} \\ 0 \end{bmatrix} \tag{2.82}$$

$$\check{P}_{k-L} = \begin{bmatrix} P_{k-L} & 0 \\ 0 & \infty \end{bmatrix} \tag{2.83}$$

where $\hat{x}_{K-L}$ is the one-step prediction estimate, and P_{k-L} is the state's prediction error covariance at the beginning of the time window $[k-L, k]$. Recall that, when the disturbance is Gaussian, the Kalman filter gives the maximum a posteriori probability or MAP estimate of the state (Sections 2.2 and A.4). The maximum likelihood estimate (MLE) of a parameter is a special case of its MAP when there is no a priori information on the parameter. Setting the initial covariance of the unknown constant ν to infinity in Eq.(2.83), therefore, gives us its MLE, $\hat{\nu}_k$.

The above method requires the starting of a new filter (and the stopping of an old one) at each time k, just as the method based on the innovation requires a sliding window. Running $k - L + 1$ simultaneous filters certainly requires more computations than running one filter and a sliding window, but the method has its advantages, such as the availability of the failure estimate $\hat{\nu}_k$ and the state estimate $\hat{x}_k$ immediately upon the failure's occurrence, i.e., before it is even detected. Another advantage is the fact that the failure model used above can be easily generalized to a Gauss Markov process with a certain bandwidth and amplitude. A priori knowledge about the failure can also be incorporated in the initial covariance matrix. If a Gauss Markov model is used for the failure, then one steady-state filter can be used to approximate the failure estimate at each time k. A Gauss Markov model is used in Chapter 5, as it can make the FDI algorithm robust to failure mode uncertainty, in addition to our objective of achieving robustness to plant model uncertainties.

From the above discussion, we see that the Kalman filter can be used in two different ways by the GLRT. It can be a whitening filter, meaning that it produces the innovations from the observations as shown in Eqs.(2.64a-2.64b). The fact that the innovations are statistically independent makes it possible to compute the likelihood ratio recursively, thus making on line computation feasible, or even easy. Another way of implementing the GLRT is to use a Kalman filter design based on the system of Eqs.(2.80-2.83). This design includes the failure as a state. Both approaches can be implemented in a way that makes them equivalent. The advantage of the first is computational: one filter continuously produces the innovations that are used in computing the likelihood ratio. Once a failure occurs, however, the Kalman filter based on Eqs.(2.64a-2.64b) no longer gives an accurate estimate of the state, but the one based on Eqs.(2.80-2.83) does.

We now discuss the isolation problem, where the objective is to determine which component failed. If there are M additive (sensors, actuators) failures under consideration, then it is possible to pose the following M hypothesis tests in terms of the innovations

$$H_0 \;\; : \;\; \varrho_k = \varrho_{ok} \tag{2.84}$$

$$H_j \;\; : \;\; \varrho_k = \varrho_{ok} + G_{j,k}\nu \tag{2.85}$$

for $j = 1, 2, ..., M$. At each time step k, each of these tests will produce a likelihood function $\ell_{j,k}$. The most likely component to have failed, j^*, is the one that produces the maximum value of the likelihood ratio. Specifically,

$$j^* = \arg\max_j \ell_{j,k} \tag{2.86}$$

and a failure in component j^* is declared if $\ell_{j^*,k}$ crosses the threshold.

2.3.2   Nonadditive Failures

Rather than assume that the failure is additive, as in Eqs.(2.58a-2.58d), we now assume that a failure causes the plant to adopt different dynamics alltogether. Specifically, we have the following pairs of hypothesis tests

$$H_0 \quad : \quad x_{k+1} \;=\; A_k x_k + B_k r_k + U_k u_k \qquad (2.87\text{a})$$

$$y_k \;=\; C_k x_k + D_k r_k + W_k u_k \qquad (2.87\text{b})$$

$$H_j \quad : \quad x_{k+1} \;=\; A_{j,k} x_k + B_{j,k} r_k + U_{j,k} u_k \qquad (2.87\text{c})$$

$$y_k \;=\; C_{j,k} x_k + D_{j,k} r_k + W_{j,k} u_k \qquad (2.87\text{d})$$

for $j = 1, ..., M$. The assumptions on the disturbance are the same as before, i.e., zero-mean, white, and Gaussian with a unit covariance matrix. Nonadditive failures occur, for instance, when the dynamics of the plant change.

In theory, the GLRT is an appropriate test for this problem. In practice, however, it is not. Recall that the key to a recursive implementation of the GLRT for additive failures is the fact that a Kalman filter design based on an unfailed plant produces the innovation even if a failure occurs. That is, we have a white residual whether a failure occurs or not. For nonadditive failures, however, a Kalman filter design based on an unfailed plant will not produce the innovation should a failure occur. As a result, the multiplicative separation of Eq.(2.67) will not be possible, and the test cannot be implemented in a recursive manner.

An alternative is to design a Kalman filter based on the model provided by each of the hypothesis. This is one version of the Multiple Model method (See also [116]). We therefore have M *conditionally* white residuals, and we can compute the likelihood ratio test, and select the most likely model.

The trouble with the Multiple Model method is that the number of possible failures may be too large for online implementation. If, in addition, the time of the hypothesis is to be taken into account, then it can be shown that the Multiple Model method requires a bank of Kalman filters that grows with time.

We shall not discuss nonadditive failures any further, as a robust extension to the multiple model method is still an open research problem.

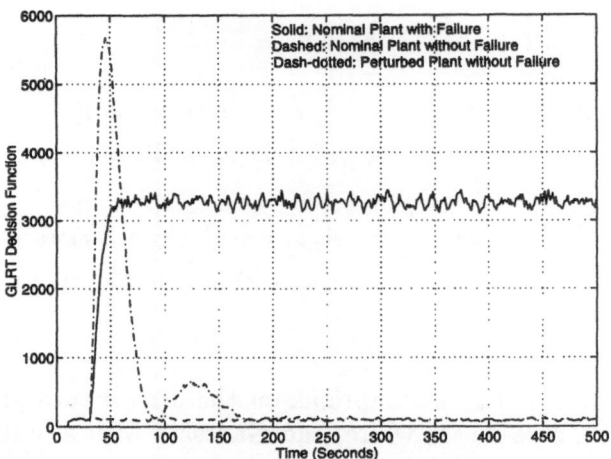

Figure 2.7: Effect of a maneuver on the GLRT's decision function for nominal and perturbed plants.

2.3.3   Modeling Uncertainties and FDI Algorithms

A limitation of the GLRT algorithm, as well as of many other FDI schemes, is the reliance on a perfect model of the plant and sensors. Any deviation from the model assumptions, such as unmodeled dynamics, parameter perturbations, or noise statistical distribution uncertainty, can lead to a serious degradation in performance of these algorithms [79]. We illustrate this point with an example.

Consider an additive perturbation where the matrix U_k in Eq.(2.80) is actually $U_k + \Delta U$, and an input command u_k is present. As before, the Kalman filter design is based on the nominal system. To demonstrate the effect of additive model uncertainties on the performance of the GLRT, a test was designed for the detection of step failures in the fins of an underwater vehicle (Sections 6.2 and D.1). For this application, a pitch maneuver starts at time $t = 30$ seconds. The horizontal fins are to be deflected 10 degrees and to remain there for at least 500 seconds. The failure occurs at the start of the maneuver, when the fin is stuck at 5 degrees.

The plots of Figure 2.7 represent a simulation of the behavior of the GLRT decision function as a function of time. The GLRT window size in this run was 5 seconds. In the absence of a failure and any model uncertainty, the GLRT remains at or near zero level, in spite of the maneuver (dashed curve). In the presence of a failure and with accurate knowledge of the plant model, the figure shows that the GLRT decision function rises quickly after the failure occurs at time $t = 30$ seconds (solid curve). The maneuver did not affect the GLRT decision function, as the Kalman filter properly took it into account. Finally, in the presence of model uncertainties, but with no failure occurring, the figure shows that the maneuver causes the GLRT decision function to rise drastically (dash-dotted curve). The decision function does eventually settle down, but only after time $t = 150$ seconds.

This experiment shows that model uncertainty can totally disable the GLRT, as it is not possible to select a threshold level for the above combination of maneuver and failure. Any choice of threshold level is either too low to avoid a false alarm in the presence of model uncertainty, or too high to allow for the failure to be detected shortly after it occurs in the nominal system. Other runs with different window lengths gave similar results. A discussion on the effect of model uncertainty on the GLRT's performance is found in Chapter 5. We will see more of this undersea vehicle application in Chapters 5 and 6.

2.3.4   Robust Failure Detection and Isolation

A robust FDI algorithm can account for several categories of modeling uncertainties. These include : Noise and plant model uncertainties, failure mode uncertainties, and statistical outliers (Eq.(2.55)). We will be concerned with the first two kinds of uncertainties.

The likelihood ratio test can be generalized to muliple mode failures. In ([42], Page 338), for instance, a general failure model of the form

$$f_k = \sum_{i=1}^{L} \phi_{ik} \alpha_i \qquad (2.88)$$

is suggested for use in a likelihood ratio context. Here, the ϕ_{ik}'s span a subspace of functions. Hall [47] replaced the step failure model of the GLRT with exactly such a model to obtain an algorithm that is robust to failure mode uncertainty.

Work on detection methods that are robust to statistical outliers is found in [66], and references therein. The focus of the work in [66] is on general signal detection, however, and not on failure detection for dynamic systems. Failure detection algorithms in dynamic systems that are robust to statistical outliers have yet to be developed. These algorithms could make use of the filters derived in [92].

Previous work on robustness to plant model uncertainty includes that of Lou, Willsky, and Verghese [76], where a geometric interpretation of the concept of analytical redundancy leads to a procedure involving singular value decompositions for determining redundancy relations that are maximally insensitive to model uncertainties. Gertler and Singer [36] use an alternative approach, where they assume that model errors may be deduced from the uncertainties of a set of underlying parameters. The partial derivatives of the residuals with respect to these parameters are then computed and the residual generator with lowest partial sensitivity is selected. Another approach is that of Horak [52], Emami-Naeini, Akhter, and Rock [28], and Tsui [107], where a bound on the effect of model uncertainties on the residual is estimated. This bound is then used to set the threshold accordingly. Work based on the unknown input observer of Frank and Patton is found in the book [87], and in other references by the same authors. Other work based on the unknown input observer is that of Guan and Saif [46]. The unknown input observer attempts to decouple failures from unknown inputs that are possibly due to noise and model uncertainties. Both Frank and Patton contributed significantly to the problem of robust detection using different techniques. The use of interrogative inputs to systems in order to robustly detect and isolate failures has been studied by Riggins [91], and Ribbens and Riggins [90].

Previous work using H_∞ techniques include the work of Frank and Ding [32], where a steady-state frequency domain based filter design is used to attenuate the effect of disturbances. In [79], a preliminary study by Mangoubi, Appleby, and Farrell on the use of H_∞ filters with robustness to noise and plant model uncertainties for failure detection shows that these filters can generate output residuals that are highly insensitive to these uncertainties. In [27], a failure detection scheme that makes use of an H_∞ filter and its output residual is proposed. The detection method in [27], however, is not robust to plant model uncertainties, as the H_∞ filter used assumes an accurate plant model.

The approach of Chapter 5 uses a general first-order Gauss Markov model for the failure to obtain robustness to failure mode uncertainties. This Gauss Markov model is appended to the plant dynamic model, and a robust H_∞ or risk sensitive estimator, such as those developed in Chapters 3 and 4, is used for synthesizing residuals that are robust to failure mode, noise and plant model uncertainties.

2.4 Summary

This chapter presents an overview of estimation and failure detection for dynamic systems. The Wiener and Kalman filters, the most commonly used linear observers, are described. These filters assume a linear plant whose model is accurate and disturbances whose first two moments are known. They provide the *linear* least squares or minimum error variance estimate of the plant's states. Their sensitivity to modeling uncertainties was demonstrated, thus motivating the robust H_∞ and risk sensitive estimation algorithms of Chapters 3 and 4, and Appendix C. Previous work on robust filtering is also discussed. A tutorial on the Kalman filter is given in Appendix A, and the filter's equations are summarized in Section A.10.

The role of linear filters in failure detection and isolation for dynamic systems was also described. The GLRT, which uses the Kalman filter, was derived. It is shown that the GLRT can use the Kalman filter either as a whitening filter, or as a maximum likelihood estimator for the failure. Examples are used to demonstrate the sensitivity of the GLRT and the detection filter to model uncertainties, and to motivate the robust failure detection and isolation algorithm to be derived in Chapter 5.

CHAPTER 3
DISCRETE-TIME ROBUST ESTIMATION

3.1 Introduction

In this chapter we derive estimators for discrete-time linear systems that are robust to plant and noise model uncertainties. The approach used is based on a game theoretic formulation in which the disturbances and the modeling errors act as opponents of the state estimator. Specifically, it is assumed that an adversarial player is attempting to manipulate, within specified constraints, the initial condition, the disturbances, and the plant so as to maximize the 2-norm of the state estimation error, which we are trying to minimize. The goal of the estimator is therefore to find the best state estimate — in the least squares sense — for the *worst* possible combination of initial condition, disturbances and modeling errors.

In Section 3.2, an estimator is derived for time-varying finite-horizon discrete-time linear systems where the plant model dynamics are assumed to be accurate, but where the models of the disturbances — process and measurement noise as well as initial state value — are unknown. The only assumption made in the derivation is that the noise has a bounded 2-norm. This estimator is robust in the sense that the estimation error satisfies a norm bound for the entire class of norm bounded disturbances and initial errors considered. Finally, it is shown that the game theoretic estimator is a generalization of both the Kalman filter and the *induced* 2-norm or H_∞ estimator. In fact, we derive an entire family of estimators that permits the designer to tradeoff least squares error performance and robustness to noise.

In Section 3.3, estimators are derived for finite-horizon discrete-time linear systems where both the plant dynamics and the disturbance models are uncertain. The assumptions for the disturbances are the same as in Section 3.2, i.e., they are norm bounded. The only assumption on the plant perturbations is that they have a bounded gain or induced 2-norm. The solution is obtained in two steps, with each step requiring the solution of a Riccati equation. In the first step, a completion of square argument is used to transform the problem into one that has the same form as the estimator obtained in Section 3.2. The second step consists of solving the transformed problem, which requires the solution to a second Riccati equation.

In Section 3.4, we present the solution to the steady-state infinite-horizon problem for stable time-invariant systems. In steady state, minimizing the induced 2-norm is the same as minimizing the H_∞ norm of the transfer function between the disturbance and the steady-state error. The H_∞ norm is the maximum gain of a transfer function over all frequencies. An estimator that minimizes this norm, therefore, minimizes the error at the frequency where the magnitude of this error is largest.

The problem of fixed-interval smoothing is considered in Section 3.5. First, the smoothing problem for systems with known plant dynamics but uncertain noise model is described. Since measurements for the entire interval are available to the estimator, all the opponent's moves can be estimated. The objective of the smoother is therefore reduced to that of finding the best estimate — in the least squares sense — of the process noise and initial condition. This is no different than the case where the disturbances are assumed Gaussian with known parameters. The fixed-interval smoothing equations are therefore the same as those of the least squares, or Kalman, smoother. A smoother with robustness to plant perturbations is also presented in Section 3.5.

Section 3.6 gives two numerical examples. The first is a simple two-state example, and the second is an application of robust filters to model based attitude determination. Section 3.7 concludes the chapter.

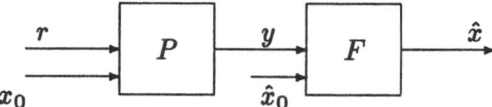

Figure 3.1: Input-Output representation of plant estimation problem.

3.2   Plants with an Uncertain Noise Model

Figure 3.1 shows an input/output representation of a nominal plant P and an estimator F. All inputs and outputs are real vectors. For $k \in [0, N-1]$, the exogenous input to the plant is the vector r_k (which includes both process and measurement noise), and the actual initial state x_0 of the plant. The inputs to the filter are the measurement and the initial state estimate, $\hat{x}_0$. Both the initial condition and the initial error $x_0 - \hat{x}_0$ are assumed to lie in a set defined by a Euclidean norm bound. The plant output is the measurement vector y_k. This output is in turn fed to the estimator F, whose output is the state estimate $\hat{x}$. The following notation will be used to denote the input and output sequences

$$r \equiv [r_0, ..., r_{N-1}]$$
$$y \equiv [y_0, ..., y_{N-1}]$$

For the state estimation error, two cases can be considered: The predicted (a priori) error, e, which is based on the one-step prediction estimate $\hat{x}_k$ that is a function of the observations up to time $k-1$,

$$e \equiv [e_1, ..., e_N]$$
$$e_k \equiv M_k(x_k - \hat{x}_k)$$

and the filtered (a posteriori) estimation error, e_+, which is based on the filtered estimate $\hat{x}_{k|k}$ that is a function of the observations up to time k,

$$e_+ \equiv \left[e_{0|0}, ..., e_{N-1|N-1}\right]$$
$$e_{k|k} \equiv M_k\left(x_k - \hat{x}_{k|k}\right)$$

Notice the interval shift for e_+. The estimator will be first derived for the one-step predictor, and we will show that results for the filtered estimate are a special case of the one-step predictor. The 2-norms of these sequences will be defined as

$$\|r\| \equiv \left(\sum_{k=0}^{N-1} r'_k r_k\right)^{1/2}$$

$$\|e\| \equiv \left(\sum_{k=1}^{N} e'_k e_k\right)^{1/2}$$

$$\|e_+\| \equiv \left(\sum_{k=0}^{N-1} e'_{k|k} e_{k|k}\right)^{1/2}$$

In addition, we denote the Euclidean norm of a vector s by $\|s\|$ and the weighted Euclidean norm by $\|s\|_Q$, where Q is a symmetric positive definite matrix of appropriate dimension.

3.2.1   Problem Formulation

The game theoretic estimator seeks to bound and minimize the induced norm of the operator from the input disturbances r and initial error $(x_0 - \hat{x}_0)$ to the state estimation prediction error e, or filtered estimation error e_+. As mentioned before, the derivation will be carried out for the predicted state estimation error only, but the results for the filtered estimation error will also be stated. If $\mathcal{G}$ is the mapping from $(r, x_0 - \hat{x}_0)$ to e, then our goal is to select a filter F that achieves a bound on $\|\mathcal{G}\|_{i2}$, the induced 2-norm of $\mathcal{G}$. Specifically,

$$\|\mathcal{G}\|_{i2}^2 \equiv \sup_{(r, x_0 - \hat{x}_0) \neq 0} \frac{\|e\|^2}{\|r\|^2 + \|x_0 - \hat{x}_0\|_{P_0^{-1}}^2}$$

$$\equiv \sup_{\{\|r\|^2 + \|x_0 - \hat{x}_0\|_{P_0^{-1}}^2 \leq 1\}} \|e\|^2$$

$$< 1 \tag{3.1}$$

The weighting matrix P_0, which is assumed to be symmetric and positive definite, is a measure of the uncertainty in the initial guess. We note that no loss of generality is incurred by setting the bound on $\|\mathcal{G}\|_{i2}$ in Eq.(3.1) to one, as the matrices B_k and D_k can be rescaled.

We assume the nominal plant has the linear time-varying state-space representation

$$x_{k+1} = A_k x_k + B_k r_k$$
$$e_k = M_k(x_k - \hat{x}_k)$$
$$y_k = C_k x_k + D_k r_k$$

$$\left[\begin{array}{c} x_{k+1} \\ \hline e_k \\ y_k \end{array} \right] = \left[\begin{array}{c|ccc} A_k & B_k & 0 \\ \hline M_k & 0 & -M_k \\ C_k & D_k & 0 \end{array} \right] \left[\begin{array}{c} x_k \\ r_k \\ \hat{x}_k \end{array} \right] \qquad (3.2)$$

with initial condition x_0 and initial error $e_0 \equiv x_0 - \hat{x}_0$. The zero entries in the last column are due to the fact that the plant state estimate does not affect the plant state or the plant output. Likewise, the zero entry in the second row is due to the fact that the input noise does not enter into the error definition.

Note that perfectly known inputs can be absorbed into such a nominal model as follows. Let u_k be the known input vector, and perform the substitutions

$$r_k \leftarrow \left[\begin{array}{c} r_k \\ u_k \end{array} \right]$$

$$y_k \leftarrow \left[\begin{array}{c} y_k \\ u_k \end{array} \right]$$

$$C_k \leftarrow \left[\begin{array}{cc} C_k & 0 \\ 0 & 0 \end{array} \right]$$

$$D_k \leftarrow \left[\begin{array}{cc} D_k & 0 \\ 0 & I \end{array} \right]$$

$$B_k \leftarrow \left[B_k \;\; E_k \right]$$

where E_k describes how the input u_k enters the system. The augmentations to the matrices C_k and D_k are of dimensions corresponding to u_k. The idea behind these substitutions is to represent u_k as a disturbance that is measured exactly.

In order to achieve the bound on $\|\mathcal{G}\|_{i2}$ in Eq.(3.1), we define a minmax or game theoretic estimation problem that minimizes an objective with respect to the state estimate $\hat{x}$ in the presence of the worst possible input r and initial state x_0. Next, we show that the estimation error obtained when using the solution to the game theoretic problem satisfies the bound on $\|\mathcal{G}\|_{i2}$ of Eq.(3.1). The game formulation is given as

$$\min_{\hat{x}} \max_{r, x_0} J_1 \qquad (3.3)$$

$$\text{subject to} \quad \text{Eq.(3.2)}$$

$$\text{and} \quad \|r\|^2 + \|x_0 - \hat{x}_0\|_{P_0^{-1}}^2 \leq 1$$

where

$$J_1 \equiv \frac{1}{2}\|e\|^2$$

$$\equiv \frac{1}{2}\sum_{k=1}^{N}(x_k - \hat{x}_k)' M_k' M_k (x_k - \hat{x}_k)$$

To solve the above game theoretic or minmax problem, we first maximize with respect to the initial condition and disturbance signal, or more specifically with respect to x_0 and $r_0, \ldots, r_{N-1}$, and then minimize with respect to the state estimate at each step, or $\hat{x}_0, \ldots, \hat{x}_{N-1}$. This means that we are implicitly assuming that the opposing player who chooses the initial error and the noise is clever enough to select the worst possible disturbance, and we must select the state estimate so as to be ready for that worst case. The order in which the players make their moves needs not be exactly as we chose. Under some conditions, the solution is the same regardless of the order. This issue is discussed in Section 3.2.3.

3.2.2   Derivation of the Estimator

We will choose the dynamics of the estimator to have the same form as the one-step Kalman predictor

$$\hat{x}_{k+1} = (A_k - K_k C_k)\hat{x}_k + K_k y_k \qquad (3.4)$$

Note that Eq.(3.4) implies that the estimator is linear. It is also unbiased in the sense that $x = \hat{x}$ in the absence of noise and initial error. The dynamics of the estimation error $\tilde{x}_k \equiv x_k - \hat{x}_k$ can then be represented as

$$\begin{aligned}
\tilde{x}_{k+1} &= (A_k - K_k C_k)\tilde{x}_k + (B_k - K_k D_k)r_k \\
&= \tilde{A}_k \tilde{x}_k + \tilde{B}_k r_k \\
e_k &= M_k \tilde{x}_k \qquad (3.5)
\end{aligned}$$

The estimation problem can now be described as

$$\min_{\{K\}} \max_{r, \tilde{x}_0} \quad J_1 \tag{3.6}$$

$$\text{given} \quad \tilde{x}_{k+1} = \tilde{A}_k \tilde{x}_k + \tilde{B}_k r_k$$

$$\text{and} \quad \|r\|^2 + \|\tilde{x}_0\|^2_{P_0^{-1}} \leq 1$$

where $\{K\} = \{K\}_{k=0}^{N-1} \equiv [K_0, ..., K_{N-1}]$. If $\lambda_1, ..., \lambda_N$ are the Lagrange multipliers associated with the dynamic constraints in Eq.(3.2), and $\frac{\gamma^2}{2}$ is associated with the disturbance norm constraint in Eq.(3.6), then the augmented cost function is given by

$$
\begin{aligned}
J_2 &\equiv J_1 + \sum_{k=0}^{N-1} \left(\lambda'_{k+1}(\tilde{x}_{k+1} - \tilde{A}_k \tilde{x}_k - \tilde{B}_k r_k) \right) - \frac{\gamma^2}{2} \left(\|r\|^2 + \|\tilde{x}_0\|^2_{P_0^{-1}} \right) \\
&= -\frac{\gamma^2}{2} \left(r'_0 r_0 + \tilde{x}'_0 P_0^{-1} \tilde{x}_0 \right) + \lambda'_1 \left(-\tilde{A}_0 \tilde{x}_0 - \tilde{B}_0 r_0 \right) \\
&\quad + \sum_{k=1}^{N-1} \left(\lambda'_k \tilde{x}_k + \frac{1}{2} \tilde{x}'_k M'_k M_k \tilde{x}_k - \frac{1}{2} \gamma^2 r'_k r_k \right. \\
&\quad \left. + \lambda'_{k+1} \left(-\tilde{A}_k \tilde{x}_k - \tilde{B}_k r_k \right) \right) \\
&\quad + \frac{1}{2} \tilde{x}'_N M'_N M_N \tilde{x}_N + \lambda'_N \tilde{x}_N
\end{aligned}
$$

Taking the variation for $k = 0, ..., N-1$, we get

$$
\begin{aligned}
\delta J_2 &= (-\gamma^2 \tilde{x}'_0 P_0^{-1} - \lambda'_1 \tilde{A}_0) \delta \tilde{x}_0 \\
&\quad + \sum_{k=1}^{N-1} \left(\tilde{x}'_k M'_k M_k - \lambda'_{k+1} \tilde{A}_k + \lambda'_k \right) \delta \tilde{x}_k \\
&\quad + \sum_{k=0}^{N-1} \left(-\gamma^2 r'_k - \lambda'_{k+1} \tilde{B}_k \right) \delta r_k \\
&\quad + \sum_{k=0}^{N-1} \left(\tilde{x}_{k+1} - \tilde{A}_k \tilde{x}_k - \tilde{B}_k r_k \right)' \delta \lambda_k
\end{aligned}
$$

Setting $\delta J_2 = 0$ gives (with $M_0 \equiv 0$)

$$
\begin{aligned}
r^*_k &= -\gamma^{-2} \tilde{B}'_k \lambda_{k+1} & \text{(3.7a)} \\
\tilde{x}_{k+1} &= \tilde{A}_k x_k + \tilde{B}_k r_k & \text{(3.7b)} \\
\tilde{x}^*_0 &= -\gamma^{-2} P_0 \tilde{A}'_0 \lambda_1 & \text{(3.7c)} \\
&= -\gamma^{-2} P_0 \lambda_0 & \text{(3.7d)} \\
-\lambda_k &= M'_k M_k \tilde{x}_k - \tilde{A}'_k \lambda_{k+1} & \text{(3.7e)}
\end{aligned}
$$

We will show below (Theorem 3.2.3) that r_k^*, $\tilde{x}_0^*$ is maximizing. Substituting for r_k^* from Eq.(3.7a) into Eq.(3.7b) gives the Hamiltonian equation

$$\begin{bmatrix} \tilde{x}_{k+1} \\ -\lambda_k \end{bmatrix} = \begin{bmatrix} \tilde{A}_k & -\gamma^{-2}\tilde{B}_k\tilde{B}_k' \\ M_k'M_k & -\tilde{A}_k' \end{bmatrix} \begin{bmatrix} \tilde{x}_k \\ \lambda_{k+1} \end{bmatrix} \qquad (3.8a)$$

$$\tilde{x}_0 = -\gamma^{-2}P_0\lambda_0 \qquad (3.8b)$$

$$\lambda_{N+1} \qquad \text{free} \qquad (3.8c)$$

If we take $r_N \equiv 0$, then $\lambda_{N+1} = 0$ is the logical guess. The associated Riccati equation can be derived by first noting that the state estimation error $\tilde{x}_k$ and the co-state λ_k are linearly related. Let Φ be the transition matrix of the Hamiltonian equation (3.8a-3.8c). Then,

$$\begin{bmatrix} \tilde{x}_k \\ \lambda_k \end{bmatrix} = \begin{bmatrix} \phi_{11}(k,0) & \phi_{12}(k,0) \\ \phi_{21}(k,0) & \phi_{22}(k,0) \end{bmatrix} \begin{bmatrix} \tilde{x}_0 \\ \lambda_0 \end{bmatrix}$$

Now,

$$\begin{aligned} 0 &= \tilde{x}_0 + \gamma^{-2}P_0\lambda_0 \\ &= \phi_{11}(0,k)\tilde{x}_k + \phi_{12}(0,k)\lambda_k + \gamma^{-2}P_0\left(\phi_{21}(0,k)\tilde{x}_k + \phi_{22}(0,k)\lambda_k\right) \\ &= \left(\phi_{11}(0,k) + \gamma^{-2}P_0\phi_{21}(0,k)\right)\tilde{x}_k \\ &\quad + \left(\phi_{12}(0,k) + \gamma^{-2}P_0\phi_{22}(0,k)\right)\lambda_k \end{aligned}$$

This implies that

$$\tilde{x}_k = -\gamma^{-2}P_k\lambda_k \qquad (3.9)$$

with

$$P_k \equiv \left(\gamma^2\phi_{11}(0,k) + P_0\phi_{21}(0,k)\right)^{-1}\left(\phi_{12}(0,k) + \gamma^{-2}P_0\phi_{22}(0,k)\right) \qquad (3.10)$$

assuming $[\gamma^2\phi_{11}(0,k) + P_0\phi_{21}(0,k)]$ is invertible. Substituting for λ_k from the Hamiltonian equation (3.8a) into Eq.(3.9), we get

$$\tilde{x}_k = \gamma^{-2}P_k\left(M_k'M_k\tilde{x}_k - \tilde{A}_k'\lambda_{k+1}\right)$$

or,

$$\tilde{x}_k = -\gamma^{-2}\left(I - \gamma^{-2}P_kM_k'M_k\right)^{-1}P_k\tilde{A}_k'\lambda_{k+1} \qquad (3.11)$$

Substituting for $\tilde{x}_k$ from the above equation into the top row of Eq.(3.8a) gives

$$\tilde{x}_{k+1} = -\gamma^{-2}\left(\tilde{A}_k\left(I - \gamma^{-2}P_kM_k'M_k\right)^{-1}P_k\tilde{A}_k' + \tilde{B}_k\tilde{B}_k'\right)\lambda_{k+1} \qquad (3.12)$$

Equating Eq.(3.9) at time $k+1$ with Eq.(3.12) gives

$$
\begin{aligned}
-\gamma^{-2} P_{k+1}\lambda_{k+1} &= -\gamma^{-2}\left(\tilde{A}_k\left(I - \gamma^{-2}P_k M_k' M_k\right)^{-1} P_k \tilde{A}_k' + \tilde{B}_k \tilde{B}_k'\right)\lambda_{k+1} \\
&= -\gamma^{-2}\left(\tilde{A}_k\left(P_k^{-1} - \gamma^{-2}M_k' M_k\right)^{-1} \tilde{A}_k' + \tilde{B}_k \tilde{B}_k'\right)\lambda_{k+1}
\end{aligned}
$$

From Eqs.(3.8a-3.8c), if $\tilde{A}_k$ is invertible, which we assume, and since λ_N is free, λ_k can take any arbitrary value. As a result, the above equation must hold true for any arbitrary value of λ_{k+1}. It follows that

$$
\begin{aligned}
P_{k+1} &= \tilde{A}_k H_k \tilde{A}_k' + \tilde{B}_k \tilde{B}_k' \quad k = 0, ..., N-1 &\text{(3.13a)} \\
P_0 &= \text{Weight on initial error} &\text{(3.13b)}
\end{aligned}
$$

where

$$
H_k \equiv \left(P_k^{-1} - \gamma^{-2}M_k' M_k\right)^{-1} \tag{3.14}
$$

From Eq.(3.10), $P_k > 0$. The parameter γ is therefore bounded below by the requirement that H_k be positive definite. Recall that $\frac{\gamma^2}{2}$ is a Lagrange multiplier. It now becomes a free parameter, as we refrain from solving the variational problem all the way to the end, since our original objective is to prove that the induced norm of the mapping from disturbance to error is bounded. We will show in the next section that a positive definite H_k permits us to demonstrate the bound's existence.

The optimal gain K_k is obtained by substituting the worst case noise and initial condition, r_k^* and $\tilde{x}_0^*$, respectively from Eqs.(3.7a,3.7d), as well as $\tilde{x}_k$ from Eq.(3.11), into the objective function J_1, and optimizing the resulting expression with respect to K_k. Thus, let

$$
J_3 = \max_{r,\tilde{x}_0} J_1
$$

Then,

$$
\begin{aligned}
J_3 &= \max_{r,\tilde{x}_0} \frac{1}{2}\sum_{k=0}^{N-1} \tilde{x}_{k+1}' M_{k+1}' M_{k+1}\tilde{x}_{k+1} \\
&= \max_{r,\tilde{x}_0} \frac{1}{2}\sum_{k=0}^{N-1} \left(\tilde{A}_k x_k + \tilde{B}_k r_k\right)' M_{k+1}' M_{k+1}\left(\tilde{A}_k x_k + \tilde{B}_k r_k\right) \\
&= \frac{1}{2\gamma^2}\sum_{k=0}^{N-1} \lambda_{k+1}'\left(\tilde{A}_k H_k \tilde{A}_k' + \tilde{B}_k \tilde{B}_k'\right)' M_{k+1}' M_{k+1}
\end{aligned}
$$

$$\left(\tilde{A}_k H_k \tilde{A}'_k + \tilde{B}_k \tilde{B}'_k \right) \lambda_{k+1}$$

$$= \frac{1}{2\gamma^2} \sum_{k=0}^{N-1} \mathrm{trace} \left((\lambda_{k+1} \lambda'_{k+1}) \left(\tilde{A}_k H_k \tilde{A}'_k + \tilde{B}_k \tilde{B}'_k \right)' M'_{k+1} M_{k+1} \right.$$

$$\left. \left(\tilde{A}_k H_k \tilde{A}'_k + \tilde{B}_k \tilde{B}'_k \right) \right)$$

Differentiating with respect to K_k, we get

$$\frac{\partial J_3}{\partial K_k} = \frac{1}{\gamma^2} \sum_{k=0}^{N-1} \mathrm{trace} \left((\lambda_{k+1} \lambda'_{k+1}) \left(\tilde{A}_k H_k \tilde{A}'_k + \tilde{B}_k \tilde{B}'_k \right)' M'_{k+1} M_{k+1} \right.$$

$$\left. \frac{\partial}{\partial K_k} \left(\tilde{A}_k H_k \tilde{A}'_k + \tilde{B}_k \tilde{B}'_k \right) \right)$$

$$= \frac{1}{\gamma^2} \sum_{k=0}^{N-1} \mathrm{trace} \left\{ (\lambda_{k+1} \lambda'_{k+1}) \left(\tilde{A}_k H_k \tilde{A}'_k + \tilde{B}_k \tilde{B}'_k \right)' M'_{k+1} M_{k+1} \right.$$

$$\frac{\partial}{\partial K_k} \left((A_k - K_k C_k) H_k (A_k - K_k C_k)' \right.$$

$$\left. \left. + (B_k - K_k D_k)(B_k - K_k D_k)' \right) \right\}$$

The above derivative vanishes for

$$K_k^* = (B_k D'_k + A_k H_k C'_k)(D_k D'_k + C_k H_k C'_k)^{-1} \tag{3.15}$$

while the Hessian with respect to K_k is

$$\frac{\partial^2 J_3}{\partial K_k^2} = \frac{1}{\gamma^2} \left((\lambda_{k+1} \lambda'_{k+1}) \frac{\partial}{\partial K_k} \left(\tilde{A}_k H_k \tilde{A}'_k + \tilde{B}_k \tilde{B}'_k \right)' M'_{k+1} M_{k+1} \right.$$

$$\left. \left(\frac{\partial}{\partial K_k} \left(\tilde{A}_k H_k \tilde{A}'_k + \tilde{B}_k \tilde{B}'_k \right) + (C_k H_k C'_k + D_k D'_k) \right) \right)$$

$$\tag{3.16}$$

If we assume that $C_k H_k C'_k + D_k D'_k$ is positive definite, then the above Hessian is also positive definite, which ensures a global minimum since J_3 is quadratic. The requirement that $C_k H_k C'_k + D_k D'_k$ be positive definite is satisfied whenever the measurements noise is independent, or, equivalently, whenever $D_k D'_k$ is invertible. If C_k is of full row rank, then the requirement on $D_k D'_k$ is not needed. In the important case of systems with redundant measurement sensors for the purpose of fault tolerance, the rows of C_k are not independent, so that the requirement on D_k is needed.

The a priori or one-step predictor estimator is therefore given by Eq.(3.4), with gain K_k given by Eq.(3.15) in terms of the Riccati matrix satisfying Eqs.(3.13a, 3.13b). Before studying the estimator properties, we give the results for the filtered estimate $\hat{x}_{k|k}$. It is easy to see that measurement update is a special case of one-step prediction. Specifically, introduce a fictitious intermediate step, $k|k$, between times k and $k+1$, when the plant is static, so that $A_k \equiv I$, and there is no process noise, so that $B_k \equiv 0$. We then have

$$
\begin{aligned}
\hat{x}_{k|k} &= (I - L_k C_k)\,\hat{x}_k + L_k y_k \\
&= \hat{x}_k + L_k\,(y_k - C_k\hat{x}_k) \quad\quad\quad\quad\text{(3.17a)} \\
L_k &= H_k C_k'\,(C_k H_k C_k' + D_k D_k')^{-1} \quad\quad\quad\text{(3.17b)}
\end{aligned}
$$

where H_k and P_k are the same as for the one-step predictor.

3.2.3   Estimator Properties

We now describe some of the properties of the estimator. We first show that the estimator achieves the desired bound of Eq.(3.1) provided the matrices P_k of the Riccati equation are positive definite with $\gamma \le 1$. We also show that the solution $(\{K\}^*, (r^*, x_0^*))$ is a game saddle point[1]. We will also mention the implications of the saddle point property.

Theorem 3.1 $\|\mathcal{G}\|_{i2}^2 < 1$ *if and only if $H_k > 0$ (and hence $P_k > 0$), for $\gamma \le 1$.*

Proof: Sufficiency. Notice that $\|\mathcal{G}\|_{i2}^2 < \gamma$ is equivalent to $\mathcal{J} < 0 \ \forall\ r,\ x_0$, where

$$
\mathcal{J} = \frac{1}{2}\|e\|^2 - \frac{\gamma^2}{2}\left(\|r\|^2 + \|x_0 - \hat{x}_0\|_{P_0^{-1}}^2\right)
$$

Inverting Eq.(3.13a), P_k^{-1} satisfies the backward recursion

$$
P_k^{-1} = \tilde{A}_k' \mathcal{V}_k^{-1} \tilde{A}_k + \gamma^{-2} M_k' M_k \quad\quad\quad\quad\text{(3.18)}
$$

where

$$
\mathcal{V}_k \equiv P_{k+1} - \tilde{B}_k \tilde{B}_k'
$$

---

[1]For a minmax or game problem $\min_\alpha \max_\beta f(\alpha, \beta)$, the pair (α^*, β^*) is a saddle point if $f(\alpha^*, \beta) \le f(\alpha^*, \beta^*) \le f(\alpha, \beta^*)$

Note that if $H_k > 0$, then $\mathcal{V}_k > 0$. Adding the identically zero term

$$\frac{\gamma^2}{2}\bigg\{ \sum_{k=0}^{N-1}(\tilde{x}'_{k+1}P_{k+1}^{-1}\tilde{x}_{k+1} - \tilde{x}'_k P_k^{-1}x_k)$$

$$+\tilde{x}'_0 P_0^{-1}\tilde{x}_0 - \tilde{x}'_N P_N^{-1}\tilde{x}_N \bigg\}$$

to $\mathcal{J}$, we get, after assembling terms,

$$\begin{aligned}
\mathcal{J} &= \frac{1}{2}\sum_{k=1}^{N}\tilde{x}'_k M'_k M_k \tilde{x}_k - \frac{\gamma^2}{2}\bigg\{\sum_{k=0}^{N-1} r'_k r_k + \tilde{x}'_0 P_0^{-1}\tilde{x}_0\bigg\} \\
&\quad +\frac{\gamma^2}{2}\bigg\{\sum_{k=0}^{N-1}(\tilde{x}'_{k+1}P_{k+1}^{-1}\tilde{x}_{k+1} - \tilde{x}'_k P_k^{-1}\tilde{x}_k) \\
&\qquad\qquad -\tilde{x}'_N P_N^{-1}\tilde{x}_N + \tilde{x}'_0 P_0^{-1}\tilde{x}_0\bigg\} \\
&= \frac{1}{2}\sum_{k=1}^{N-1}\tilde{x}'_k M'_k M_k \tilde{x}_k + \frac{1}{2}\tilde{x}'_N M'_N M_N \tilde{x}_N \\
&\quad +\frac{\gamma^2}{2}\bigg\{\sum_{k=0}^{N-1} -r'_k r_k + (\tilde{A}_k\tilde{x}_k + \tilde{B}_k r_k)' P_{k+1}^{-1}(\tilde{A}_k\tilde{x}_k + \tilde{B}_k r_k) \\
&\qquad -\tilde{x}'_k\left(\tilde{A}'_k(P_{k+1} - \tilde{B}_k\tilde{B}'_k)^{-1}\tilde{A}_k + \gamma^{-2}M'_k M_k\right)\tilde{x}_k \\
&\qquad\qquad -\tilde{x}'_N P_N^{-1}\tilde{x}_N\bigg\} \\
&= \frac{1}{2}\tilde{x}'_N(M'_N M_N - \gamma^2 P_N^{-1})\tilde{x}_N \\
&\quad +\frac{\gamma^2}{2}\bigg\{\sum_{k=0}^{N-1} -r'_k r_k + \tilde{x}'_k\tilde{A}'_k P_{k+1}^{-1}\tilde{A}_k\tilde{x}_k + 2\tilde{x}'_k\tilde{A}'_k P_{k+1}^{-1}\tilde{B}_k r_k \\
&\quad +r'_k\tilde{B}'_k P_{k+1}^{-1}\tilde{B}_k r_k - \tilde{x}'_k\tilde{A}'_k P_{k+1}^{-1}\tilde{A}_k\tilde{x}_k \\
&\quad -\tilde{x}'_k\tilde{A}'_k P_{k+1}^{-1}\tilde{B}_k\left(I - \tilde{B}'_k P_{k+1}^{-1}\tilde{B}_k\right)^{-1}\tilde{B}'_k P_{k+1}^{-1}\tilde{A}_k\tilde{x}_k\bigg\}
\end{aligned}$$

We need to eliminate $\tilde{x}_k$ from the expression for $\mathcal{J}$. Note that

$$\begin{aligned}
\tilde{B}'_k P_{k+1}^{-1}\tilde{A}_k\tilde{x}_k &= \tilde{B}'_k P_{k+1}^{-1}\left(\tilde{x}_{k+1} - \tilde{B}_k r_k\right) \\
&= r^*_k - \tilde{B}'_k P_{k+1}^{-1}\tilde{B}_k r_k
\end{aligned}$$

Substituting for $\tilde{B}'_k P_{k+1}^{-1}\tilde{A}_k\tilde{x}_k$ into the expression for $\mathcal{J}$ and defining

$$W_k \equiv \tilde{B}'_k P_{k+1}^{-1}\tilde{B}_k$$

we get

$$
\begin{aligned}
\mathcal{J} &= -\frac{\gamma^2}{2}\Bigg\{ \sum_{k=0}^{N-1} r_k' r_k - 2(r_k^* - W_k r_k)' r_k - r_k' W_{\tilde{k}} r_k \\
&\quad + (r_k^* - W_k r_k)'\,(I - W_k)^{-1}\,(r_k^* - W_k r_k) + \tilde{x}_N' H_N^{-1} \tilde{x}_N \Bigg\} \\
&= -\frac{\gamma^2}{2}\Bigg\{ \sum_{k=0}^{N-1} r_k'\left(I + W_k + W_k\,(I - W_k)^{-1}\,W_k\right) r_k \\
&\quad - 2 r_k'\left(I + W_k(I - W_k)^{-1}\right) r_k^* + r_k'^*\,(I - W_k)^{-1}\, r_k^* + \tilde{x}_N' H_N^{-1} \tilde{x}_N \Bigg\}
\end{aligned}
$$

Note that

$$
\begin{aligned}
I + W_k + W_k\,(I - W_k)^{-1}\,W_k &= (I - W_k)^{-1} \\
I + W_k\,(I - W_k)^{-1} &= (I - W_k)^{-1}
\end{aligned}
$$

Also, since $\mathcal{V}_k > 0$, we have $I - W_k > 0$. The above expression for $\mathcal{J}$ therefore reduces to

$$
\begin{aligned}
\mathcal{J} &= -\frac{\gamma^2}{2} \sum_{k=0}^{N-1} \left((r_k - r_k^*)'\,(I - W_k)^{-1}\,(r_k - r_k^*) \right) \\
&\quad - \frac{\gamma^2}{2} \tilde{x}_N' H_N^{-1} \tilde{x}_N \\
&< 0
\end{aligned} \tag{3.19}
$$

Necessity. Suppose that k^* is the smallest time step by which H_k develops a nonpositive eigenvalue. Adding the identically zero term

$$
\frac{\gamma^2}{2}\left(\sum_{k=0}^{k=k^*-1} \left(\tilde{x}_{k+1}' P_{k+1}^{-1} \tilde{x}_{k+1} - \tilde{x}_k' P_k^{-1} x_k \right) - \tilde{x}_{k^*}' P_{k^*}^{-1} x_{k^*} + \tilde{x}_0' P_0^{-1} \tilde{x}_0 \right)
$$

to $\mathcal{J}$ gives

$$
\begin{aligned}
\mathcal{J} &= -\frac{\gamma^2}{2} \sum_{k=0}^{k^*-1} (r_k - r_k^*)'\,(I - W_k)^{-1}\,(r_k - r_k^*) - \frac{\gamma^2}{2} \tilde{x}_{k^*}' H_{k^*}^{-1} \tilde{x}_{k^*} \\
&\quad + \sum_{k=k^*+1}^{N} \left(\tilde{x}_k' M_k' M_k \tilde{x}_k - \frac{\gamma^2}{2} r_k' r_k \right)
\end{aligned}
$$

Choose $r_k = r_k^*$ for $k \leq k^* - 1$ so the first term vanishes. Since H_k has a nonpositive eigenvalue, choose x_0 such that the term $\tilde{x}_{k^*}' H_{k^*}^{-1} \tilde{x}_{k^*}$ is nonpositive (This is possible if A_k is invertible, which we assume). Finally, choose

$r_k = 0$ for $k^* + 1 \leq k \leq (N - 1)$. This implies that there exists a set of input disturbances $(x_0, r) \neq 0$ such that $\mathcal{J} \geq 0$, which is a contradiction. $\diamond$

The above theorem, together with Eq.(3.16), shows that the solution to our game problem, namely $\{K\}^* = \left\{K_0^*, \ldots, K_{N-1}^*\right\}$, x_0^*, and $r^* = \left[r_0^*, \ldots, r_{N-1}^*\right]$ form a saddle point.

Theorem 3.2 *The stationary values* $(\{K\}^*, (r^*, x_0^*))$, $k = 0, \ldots, N - 1$ *satisfy*

$$J_1\left(\{K\}^*, r_k, x_0\right) \leq J_1\left(\{K\}^*, r^*, x_0^*\right) \leq J_1\left(\{K\}, r^*, x_0^*\right) \qquad (3.20)$$

Proof: The second inequality holds by virtue of Eq.(3.16), which, together with the fact that J_1 is quadratic, implies that K_k^* is a global minimum when $r_k = r_k^*$. Eq. (3.19) implies the first inequality. This is because

$$\begin{aligned}
\mathcal{J} = & -\frac{\gamma^2}{2} \sum_{k=0}^{N-1} \left((r_k - r_k^*)' \mathcal{V}_k^{-1} (r_k - r_k^*)\right) \\
& - \left(\tilde{\Phi}(N, 0)\tilde{x}_0 + \sum_{k=0}^{N-1} \tilde{\Phi}(N, k)B_k r_k^*\right)' H_N^{-1} \\
& \qquad\qquad\qquad \left(\tilde{\Phi}(N, 0)\tilde{x}_0 + \sum_{k=0}^{N-1} \tilde{\Phi}(N, k)B_k r_k^*\right)
\end{aligned}$$

where $\tilde{\Phi}$ is the transition matrix for the error dynamics . It is clear that $\mathcal{J}$ is the negative of a sum of quadratic terms in $r_k, \tilde{x}_0$. Therefore, $\mathcal{J}$ is concave, and $x_0^*, r_k^*, \forall k = 0, \ldots, N - 1$, are a global maximum for $\mathcal{J}$. Therefore they are a global maximum for J_1. $\diamond$

That the solution to our game problem satisfies the saddle point property has interesting implications. One implication is that the solution represents the optimal strategy that each of the two players can select. Another implication is that the optimal strategy for each player is the same regardless of the order of optimization [18]. Thus, we could have carried the minimization first, and the answer would not change. Alternatively, we could have required each player to make a move at each time step, and the optimal strategy would remain the same.

3.2.4   Estimator Equations and Discussion

The equations for the one-step minmax or game theoretic predictor are now summarized as

$$\hat{x}_{k+1} = (A_k - K_k C_k)\hat{x}_k + K_k y_k \tag{3.21}$$

$$K_k = (A_k H_k C_k' + B_k D_k')(C_k H_k C_k' + D_k D_k')^{-1} \tag{3.22}$$

$$P_{k+1} = (A_k - K_k C_k)H_k(A_k - K_k C_k)' \\ + (B_k - K_k D_k)(B_k - K_k D_k)' \tag{3.23}$$

$$H_k \equiv \left(P_k^{-1} - \gamma^{-2}M_k'M_k\right)^{-1} \tag{3.24}$$

with P_0 given by the weight on the initial estimation error. The filtered estimate, $\hat{x}_{k|k}$, can be obtained by setting $A_k = I$ and $B_k = 0$ in the estimator and gain equations above, as is done in Eqs.(3.17a-3.17b).

We conclude this section by discussing the meaning of the parameter γ and the relationship between the game theoretic estimator and the Kalman filter. For the purpose of this discussion, we note that, by appropriate rescaling of the matrices B_k and D_k, for $k \in [0, N-1]$, we can formulate our performance criterion as

$$\|\mathcal{G}\|_{i2} < \gamma \tag{3.25}$$

where, again, $\mathcal{G}$ is the mapping between $(x_0 - \hat{x}_0, r)$ and the error e.

The Riccati equation (3.23) and the gain (3.22) are the same as those of the Kalman filter except that the term $H_k \equiv \left(P_k^{-1} - \gamma^{-2}M_k'M_k\right)^{-1}$ replaces P_k. Recall that, in the case of the Kalman filter, P_k^{-1} is a measure of the information about the state available at time k prior to taking a measurement. Subtracting $\gamma^{-2}M_k'M_k$ from P_k^{-1} means that we believe we have less information available. As such, the smaller the choice of γ, the more conservative we are in believing our noise model. This makes the estimator robust to noise model uncertainties. The parameter γ is bounded below by the value that causes H_k to be singular since we cannot subtract more information than we have. At the lowest possible value of γ, which is found by iteration only, our estimator minimizes the induced 2-norm (H_∞) of the mapping from $(x_0 - \hat{x}_0, r)$ to the error e, defined by Eq.(3.1).

As γ increases from its minimum value, then $\gamma^{-2}M_k'M_k$ decreases. This means that we are less conservative and believe our noise model

more. But then our estimator is less robust. However, it provides
more performance with respect to the least squares error objective.
In fact, if $\gamma \to \infty$, then $H_k \to P_k$ in both the Riccati equation and
the optimal gain. In that case, one recovers the Kalman filter. The
game theoretic estimator can therefore be viewed as an extension of the
Kalman filter where decreasing the design parameter γ sacrifices nominal performance in the minimum variance sense to provide robustness
to disturbance modeling error.

Another point of interest is the role of the weighting matrix M_k.
This matrix is an additional design parameter that allows the user to
put emphasis on some variables at the expense of others. This degree
of freedom does not exist with the Kalman filter, since as $\gamma \to \infty$,
the terms in M_k disappear. On the other hand, one can argue that
the Kalman filter is always the optimal linear least squares estimator
regardless of the choice of M_k's.

3.3   Plants with Uncertain Dynamics and Noise Model

In this section, we generalize the results obtained thus far by introduc-
ing uncertainties in the model dynamics. The goal here is to minimize
the estimation error for an entire family of input initial conditions, disturbances, and plants. This is in contrast to the estimator derived in
the previous section, which assumes accurate knowledge of the plant,
and the Kalman filter, which assumes that both the plant and disturbance models are known.

Figure 3.2 shows a general input/output representation of a nominal plant P with modeling uncertainties Δ and an estimator F. The
vectors x_0, $\hat{x}_0$, r, y, and $\hat{x}$ have the same definitions as before, while
ϵ and η represent the signals connecting the nominal plant and the
perturbation

$$\eta \equiv [\eta_0, ..., \eta_{N-1}]$$
$$\epsilon \equiv [\epsilon_0, ..., \epsilon_{N-1}]$$

The 2-norms of these sequences will be denoted by

$$\|\eta\| \equiv \left(\sum_{k=0}^{N-1} \eta_k' \eta_k \right)^{1/2}$$

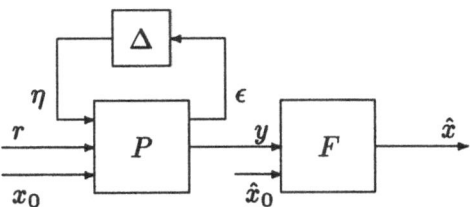

Figure 3.2: General representation of robust estimation problem

$$\|\epsilon\| \equiv \left(\sum_{k=0}^{N-1} \epsilon_k' \epsilon_k\right)^{1/2}$$

3.3.1   Problem Formulation

As with the nominal plant case, we will be concerned with the one-step predictor. The sequences r, η, ϵ, and y, are therefore defined over the interval $[0, N-1]$, while the output error e is defined over the interval $[1, N]$. The robust estimator seeks to bound the induced norm of the operator from the input disturbances r and initial estimation error $(x_0 - \hat{x}_0)$ to the estimation error e, for all possible model perturbations Δ of bounded induced 2-norm. Mathematically, we would like to satisfy the following performance criterion

$$\|\overline{\mathcal{G}}\|_{i2}^2 \equiv \sup_{\{\|r\|^2 + \|x_0 - \hat{x}_0\|_{P_0^{-1}}^2 + \|x_0\|_{\overline{X}_0}^2 = 1\}} \|e\|^2 < 1$$

$$\forall \Delta \ni \|\Delta\|_{i2}^2 \equiv \sup_{\|\epsilon\| \neq 0} \frac{\|\eta\|^2}{\|\epsilon\|^2} < 1 \tag{3.26}$$

The weighting matrix P_0, which is assumed to be symmetric and positive definite, is a measure of the uncertainty in the initial estimate. As in the previous section, the bounds are set to one by rescaling.

The approach used to achieve the performance goal of Eq.(3.26) consists of treating the perturbation output η as an additional exogenous disturbance input to the nominal plant P, and the perturbation input ϵ as an additional error term. We then find an estimator whose objective is to bound the induced norm of the mapping from the augmented input to the augmented output by one. This will guarantee that the performance objective of Eq.(3.26) is achieved. A new performance criterion is therefore defined as

$$\overline{J}_1 < 1 \qquad (3.27)$$

where

$$\overline{J}_1 \equiv \sup_{(r,\eta,x_0,x_0-\hat{x}_0)\neq 0} \frac{\|e\|^2 + \|\epsilon\|^2}{\|r\|^2 + \|\eta\|^2 + \|x_0\|^2_{\overline{X}_0} + \|x_0 - \hat{x}_0\|^2_{P_0^{-1}}}$$

The criterion of Eq.(3.27) is a robust performance or small gain condition (Sections 2.2.4-2.2.5). It is easy to show, as is done in Proposition 3.3 below, that if this criterion is satisfied, then the original performance condition of Eq.(3.26) is also satisfied for the entire class of perturbations Δ whose induced 2-norm is bounded by one.

Proposition 3.3 *If 1)* $\overline{J}_1 < 1$, *2)* $\|\Delta\|^2 < 1$, *then* $\|\overline{\mathcal{G}}\|^2_{i2} < 1$.

Proof: $\overline{J}_1 < 1$ implies that

$$\|e\|^2 + \|\epsilon\|^2 < \|r\|^2 + \|\eta\|^2 + \|x_0\|^2_{\overline{X}_0} + \|x_0 - \hat{x}_0\|^2_{P_0^{-1}}$$

for all r, η, x_0, and $x_0 - \hat{x}_0$. But from the bound on the perturbation Δ, we have

$$\|\eta\|^2 < \|\epsilon\|^2 \qquad (3.28)$$

for all ϵ. Combining the preceding two equations yields

$$\|e\|^2 < \|r\|^2 + \|x_0 - \hat{x}_0\|^2_{P_0^{-1}} + \|x_0\|^2_{\overline{X}_0}$$

for all r, $x_0 - \hat{x}_0$, and x_0. We therefore have $\|\overline{\mathcal{G}}_{i2}\|^2 < 1$. $\Diamond$

Note that Proposition 3.3 provides only a sufficient condition for satisfying our performance criterion. As such, it is conservative. Now, with $d_k \equiv [r'_k \ \eta'_k]'$, we assume the nominal plant and estimator have the linear time-varying state-space representation

$$\begin{bmatrix} x_{k+1} \\ \epsilon_k \\ e_k \\ y_k \end{bmatrix} = \left[\begin{array}{c|cc} A_k & B_k & 0 \\ \hline S_k & T_k & 0 \\ M_k & 0 & -M_k \\ C_k & D_k & 0 \end{array} \right] \begin{bmatrix} x_k \\ d_k \\ \hat{x}_k \end{bmatrix} \qquad (3.29)$$

with initial condition x_0 and initial error $\tilde{x}_0 \equiv x_0 - \hat{x}_0$. Perfectly known inputs (See the substitutions following Eq.(3.2)), as well as a large class of uncertainties, including parametric uncertainties and neglected dynamics from model reduction, can be absorbed into the above model. To incorporate neglected dynamics, one would naturally have to add some additional states. The zero entries in the last column of Eq.(3.29) are due to the fact that the plant state estimate does not affect the plant dynamics. Likewise, the zero entry in the third row is due to the fact that the input noise does not enter into the error definition.

The following example from [4] will illustrate how perturbations in the system matrices A, B, C, and D can be incorporated into the formulation of Eq.(3.29).

Example:

Consider the following state-space plant with parameter errors

$$x_{k+1} = \left(A_k + \sum_{j=1}^{l} \Delta A_{kj} \delta_j \right) x_k + \left(G_k + \sum_{j=1}^{l} \Delta G_{kj} \delta_j \right) r_k$$

$$y_k = \left(C_k + \sum_{j=1}^{l} \Delta C_{kj} \delta_j \right) x_k + \left(E_k + \sum_{j=1}^{l} \Delta E_{kj} \delta_j \right) r_k$$

Each δ_j represents a parameter error that is normalized as

$$-1 < \delta_j < 1 \quad \forall j = 1, ..., l$$

For each pair (k, j), define the matrix N_{kj}, $j = 1, ..., l$, by

$$N_{kj} = \begin{bmatrix} \Delta A_{kj} & \Delta D_{kj} \\ \Delta C_{kj} & \Delta E_{kj} \end{bmatrix} \in \mathcal{R}^{(n_x + n_y) \times (n_x + n_r)}$$

where n_x, n_y, and n_r are the dimensions of the vectors x_k, y_k, and r_k, respectively. Generally, this matrix will not be of full rank since one parameter rarely affects all of the states and outputs. So, N_{kj} can be decomposed as follows

$$N_{kj} = \begin{bmatrix} Q_{kj} \\ R_{kj} \end{bmatrix} \begin{bmatrix} S_{kj} & L_{kj} \end{bmatrix}$$

where $Q_{kj} \in \mathcal{R}^{n_x \times n_{kj}}$, $R_{kj} \in \mathcal{R}^{n_y \times n_{kj}}$, $S_{kj} \in \mathcal{R}^{n_{kj} \times n_x}$, $L_{kj} \in \mathcal{R}^{n_{kj} \times n_r}$, and n_{kj} is the rank of the matrix N_{kj}. Recalling that $d_k = [\eta_k' \ r_k']'$, the state-space model of the perturbed system can be rewritten as

$$
\begin{aligned}
x_{k+1} &= \left(A_k + \sum_{j=1}^{l} Q_{kj} \delta_j I_{n_{kj}} S_{kj} \right) x_k + \left(G_k + \sum_{j=1}^{l} Q_{kj} \delta_j I_{n_{kj}} L_{kj} \right) r_k \\
&= A_k x_k + [Q_{k1} \ \dots \ Q_{1l}] \begin{bmatrix} \eta_{k1} \\ \vdots \\ \eta_{kl} \end{bmatrix} + G_k r_k \\
&= A_k x_k + Q_k \eta_k + G_k r_k \\
&= A_k x_k + B_k d_k \\
\\
y_k &= \left(C_k + \sum_{j=1}^{l} R_{kj} \delta_j I_{n_{kj}} S_{kj} \right) x_k + \left(E_k + \sum_{j=1}^{l} R_{kj} \delta_j I_{n_{kj}} L_{kj} \right) r_k \\
&= C_{kj} x_k + [R_{k1} \ \dots \ R_{kl}] \begin{bmatrix} \eta_{k1} \\ \vdots \\ \eta_{kl} \end{bmatrix} + E_k r_k \\
&= C_k x_k + R_k \eta_k + E_k r_k \\
&= C_k x_k + D_k d_k \\
\\
\epsilon_k &= \begin{bmatrix} \epsilon_{k1} \\ \vdots \\ \epsilon_{kl} \end{bmatrix} \\
&= \begin{bmatrix} S_{k1} \\ \vdots \\ S_{kl} \end{bmatrix} x_k + 0\eta_k + \begin{bmatrix} L_{k1} \\ \vdots \\ L_{kl} \end{bmatrix} r_k \\
&= S_k x_k + T_k d_k
\end{aligned}
$$

where

$$
T_k = \begin{bmatrix} 0 & L_{k1} \\ \dots & \dots \\ 0 & L_{kl} \end{bmatrix}
$$

With the matrices B_k, D_k, Q_k, R_k, S_k, and T_k obviously defined by the above three sets of equations, we have the desired form of Eq.(3.29). The relationship between η and ϵ is then

$$
\eta = \begin{bmatrix} \eta_{k1} \\ \vdots \\ \eta_{kl} \end{bmatrix}
$$

$$= \begin{bmatrix} \delta_1 & & \\ & \ddots & \\ & & \delta_l \end{bmatrix} \begin{bmatrix} \epsilon_{k1} \\ \vdots \\ \epsilon_{kl} \end{bmatrix} \qquad (3.30)$$

$$= \Delta\epsilon \qquad (3.31)$$

We have thus shown how parametric uncertainty can be represented in the form of Figure 3.2, or Eq.(3.29).   ◇

Note that Proposition 3.3 assumes a general class of perturbations, i.e., parametric, nonparametric linear, nonlinear, complex, etc. But in the above example there are only real parametric perturbations. For this example, therefore, Proposition 3.3 is conservative in the sense that it is robust to model uncertainties that do not occur. Another source of conservatism is the structure of the uncertainty. Proposition 3.3 is robust with respect to any structure of the Δ block. But, from Eq.(3.31), we see that Δ has a block diagonal structure. Ways of accounting for the block diagonal structure of Δ so as to reduce the conservatism can be found in [4, 25, 26].

In order to achieve the condition of Eq.(3.27), we define a minmax or game theoretic estimation problem that minimizes an objective with respect to the state estimate $\hat{x} = [\hat{x}_0, \ldots, \hat{x}_{N-1}]$ in the presence of the worst possible input $d = [d_0, \ldots, d_{N-1}]$ and initial state x_0

$$\min_{\hat{x}} \max_{d, x_0} \overline{J}_2$$
$$\text{subject to} \qquad \text{Eqs.(3.29)} \qquad (3.32)$$

where

$$\overline{J}_2 \equiv \frac{1}{2}\|e\|^2 + \frac{1}{2}\|\epsilon\|^2 - \frac{\gamma^2}{2}\left(\|d\|^2 + \|x_0 - \hat{x}_0\|^2_{P_0^{-1}}\right)$$

The next section presents the solution to this game problem, which in turn yields the estimator equations. It will be shown that if there exists a solution with $\gamma \leq 1$, then

$$\overline{J}_2 - \frac{\gamma^2}{2}\|x_0\|^2_{\overline{X}_0} < 0 \qquad (3.33)$$

provided $\overline{X}_0$ is bounded below by a certain Riccati matrix, to be derived shortly. This would then imply that the robust performance criterion of Eq.(3.27) is satisfied.

3.3.2   Derivation of the Estimator

The solution to the game problem in Eq.(3.32) is obtained in two stages, with each stage requiring the solution to a Riccati equation. In the first stage, we bound the terms in the objective function that are introduced for robustness but are not affected by the estimate, $\hat{x}$. The problem is thereby reduced to the simpler estimation problem of the previous section. The second stage consists of solving the simplified problem for the optimal estimate.

The first Riccati equation is obtained by completing the square [11], [75], as is shown in Theorem 3.4 below.

Theorem 3.4 *The Riccati equation*

$$
\begin{aligned}
X_k &= A_k' X_{k+1} A_k + S_k' S_k + \gamma^{-2} F_k Z_k^{-1} F_k' \\
X_N &= 0
\end{aligned}
\tag{3.34}
$$

where

$$
\begin{aligned}
F_k &\equiv S_k' T_k + A_k' X_{k+1} B_k \\
Z_k &\equiv I - \gamma^{-2} \left(T_k' T_k + B_k' X_{k+1} B_k \right)
\end{aligned}
$$

has a solution such that $Z_k > 0$, $\forall k \in [0, N-1]$, if and only if

$$
\begin{aligned}
d_k^* &\equiv \gamma^{-2} Z_k^{-1/2} F_k' x_k \\
\bar{d}_k &\equiv Z_k^{1/2} d_k - d_k^*
\end{aligned}
\tag{3.35}
$$

result in

$$
\frac{1}{2} \left(\|\epsilon\|^2 - \gamma^2 \|d\|^2 \right) = -\frac{\gamma^2}{2} \|\bar{d}\|^2 + \frac{1}{2} x_0' X_0 x_0
\tag{3.36}
$$

Proof: Sufficiency. By adding the identicaly zero term,

$$
\sum_{k=0}^{N-1} \frac{1}{2} \left(x_{k+1}' X_{k+1} x_{k+1} - x_k' X_k x_k \right) + \frac{1}{2} x_0' X_0 x_0
$$

to

$$
\frac{1}{2} \|\epsilon\|^2 - \frac{\gamma^2}{2} \|d\|^2
$$

$$
= \frac{1}{2} \sum_{k=0}^{N-1} \left(\|S_k x_k + T_k d_k\|_2^2 - \gamma^2 \|d_k\|_2^2 \right)
$$

and substituting for x_{k+1} from Eq.(3.29) and for X_k from Eq.(3.34), we get, after assembling terms,

$$
\begin{aligned}
\frac{1}{2}\|\epsilon\|^2 - \frac{\gamma^2}{2}\|d\|^2 &= -\frac{\gamma^2}{2}\sum_{k=0}^{N-1}\left(d_k' Z_k d_k - 2\gamma^{-2}x_k' F_k d_k \right. \\
&\qquad\qquad \left. +\gamma^{-4}x_k' F_k Z_k^{-1} F_k' x_k \right) \\
&\qquad +\frac{1}{2}x_0' X_0 x_0 \\
&= -\frac{\gamma^2}{2}\sum_{k=0}^{N-1}\|Z_k^{1/2}d_k - \gamma^{-2}Z_k^{-1/2}F_k' x_k\|^2 \\
&\qquad +\frac{1}{2}x_0' X_0 x_0 \\
&= -\frac{\gamma^2}{2}\|\bar{d}\|^2 + \frac{1}{2}x_0' X_0 x_0
\end{aligned}
$$

Necessity.   Assume $k = k^*$ is the largest time step such that Z_k has a nonpositive eigenvalue. Adding the identically zero term

$$
\sum_{k=k^*}^{N-1}\frac{1}{2}\left(x_{k+1}' X_{k+1} x_{k+1} - x_k' X_k x_k\right) + \frac{1}{2}x_{k^*}' X_{k^*} x_{k^*}
$$

to $\frac{1}{2}\|\epsilon\|^2 - \frac{\gamma^2}{2}\|d\|^2$ gives, after some algebra,

$$
\begin{aligned}
\frac{1}{2}\|\epsilon\|^2 - \frac{\gamma^2}{2}\|d\|^2 &= \frac{1}{2}\sum_{k=1}^{k^*-1}\epsilon_k'\epsilon_k - \frac{\gamma^2}{2}\sum_{k=0}^{k^*-1}d_k' d_k + \frac{1}{2}x_{k^*}' X_{k^*} x_{k^*} \\
&\quad -\frac{\gamma^2}{2}\sum_{k=k^*}^{N-1}\left(Z_k^{\frac{1}{2}}d_k - d_k^*\right)'\left(Z_k^{\frac{1}{2}}d_k - d_k^*\right)
\end{aligned}
$$

Choose $x_0 = 0$, $d_k = 0$, for $k = 0, ..., k^* - 1$. This will give $x_k = 0$ for $k = 0, ..., k^*$, and the the terms in the summations for $\epsilon_k'\epsilon_k$ and for $d_k' d_k$, as well as the term $\frac{1}{2}x_{k^*}' X_k x_{k^*}$, all vanish. The terms in the last summation become

$$
-\frac{\gamma^2}{2}\left(d_{k^*}' Z_{k^*} d_{k^*} + \sum_{k=k^*}^{N-1}(Z_k^{\frac{1}{2}}d_k - d_k^*)(Z_k^{\frac{1}{2}}d_k - d_k^*)'\right)
$$

Since Z_{k^*} has a nonpositive eigenvalue, it is possible to choose a $d_{k^*} \neq 0$ such that the middle term $d_{k^*}' Z_{k^*} d_{k^*}$ is nonpositive. To cancel the terms in the last summation, choose $d_k = Z_k^{-\frac{1}{2}}d_k^*$. This will result in $\frac{1}{2}\|\epsilon\|^2 - \frac{\gamma^2}{2}\|d\|^2 \geq 0$, a contradiction. $\diamond$

Remark: Note that, if the theorem applies, then the matrices X_k are necessarily positive semidefinite for $k \in [0, N]$.

For the second stage, we seek a state estimate $\hat{x}$ such that $\overline{J}_2 <$ 0. The above theorem helps us perform a change of coordinates that simplifies the optimization problem (3.32). But first we must define our estimation problem in terms of $\overline{d}$. Define a new objective function $\overline{J}_3$ as

$$
\begin{aligned}
\overline{J}_3 &\equiv \frac{1}{2}\|e\|^2 - \frac{\gamma^2}{2}\left(\|\overline{d}\|^2 + \|x_0 - \hat{x}_0\|_{P_0^{-1}}^2\right) \\
&= \overline{J}_2 - \frac{1}{2}x_0'\overline{X}_0 x_0
\end{aligned}
\tag{3.37}
$$

and solve the optimization problem of Eq.(3.32), with $\overline{J}_3$ replacing $\overline{J}_2$, subject to the constraints of Eq.(3.29). To do so, we must express the state and observation equations (3.29) in terms of $\overline{d}$. Substituting $d_k = \gamma^{-2}Z_k^{-1}F_k'x_k + Z_k^{-1/2}\overline{d}_k$ into these equations, we get

$$
\begin{aligned}
x_{k+1} &= \overline{A}_k x_k + \overline{B}_k \overline{d}_k \tag{3.38a} \\
y_k &= \overline{C}_k x_k + \overline{D}_k \overline{d}_k \tag{3.38b}
\end{aligned}
$$

where

$$
\begin{aligned}
\overline{A}_k &= A_k + \gamma^{-2}B_k Z_k^{-1}F_k' \tag{3.39a} \\
\overline{B}_k &= B_k Z_k^{-1/2} \tag{3.39b} \\
\overline{C}_k &= C_k + \gamma^{-2}D_k Z_k^{-1}F_k' \tag{3.39c} \\
\overline{D}_k &= D_k Z_k^{-1/2} \tag{3.39d}
\end{aligned}
$$

Since the term $\|\overline{d}\|^2 + \|x_0 - \hat{x}_0\|_{P_0}^2$ is bounded, the optimization problem just described is equivalent to

$$
\min_{\hat{x}} \max_{\overline{d}, x_0} \quad \frac{1}{2}\|e\|^2 - \frac{\gamma^2}{2}\left(\|\overline{d}\|^2 + \|x_0 - \hat{x}_0\|_{P_0^{-1}}^2\right) \tag{3.40}
$$

$$
\text{subject to} \quad \text{Eqs.(3.38a) and (3.38b)} \tag{3.41}
$$

But this problem is equivalent to the one solved in Section 3.2 (Eq.(3.3)). The estimator equations are therefore

$$
\hat{x}_{k+1} = \left(\overline{A}_k - K_k \overline{C}_k\right)\hat{x}_k + K_k y_k \tag{3.42}
$$

with gain

$$
K_k = \left(\overline{A}_k H_k \overline{C}_k' + \overline{B}_k \overline{D}_k'\right)\left(\overline{C}_k H_k \overline{C}_k' + \overline{D}_k \overline{D}_k'\right)^{-1} \tag{3.43}
$$

where H_k is given by

$$H_k \equiv \left(P_k^{-1} - \gamma^{-2} M_k' M_k \right)^{-1} \tag{3.44}$$

and the Riccati matrix satisfies

$$
\begin{aligned}
P_{k+1} &= \left(\overline{A}_k - K_k \overline{C}_k \right) H_k \left(\overline{A}_k - K_k \overline{C}_k \right)' \\
&\quad + \left(\overline{B}_k - K_k \overline{D}_k \right) \left(\overline{B}_k - K_k \overline{D}_k \right)' \quad k = 0, ..., N-1 \\
P_0 &\quad \text{given} \tag{3.45}
\end{aligned}
$$

The existence of positive definite matrices H_k as well as positive definite P_k's and Z_k's in the two Riccati equations guarantees the existence of a robust estimator that satisfies the norm bound of Eq.(3.26).

Theorem 3.5 *Assume that for a $\gamma \leq 1$,*

1. *There exists a X_k, necessarilly positive semidefinite, satisfying the Riccati Equation (3.34) on $[0, N]$ such that Z_k is positive definite,*

2. *There exists a $P_k > 0$ on the same interval satisfying the Riccati Equation (3.45) such that $H_k > 0$, and*

3. *$\frac{\gamma^2}{2} \overline{X}_0 \geq X_0$*

Then the robust performance condition of Eq.(3.26) is satisfied, namely, $\|\overline{\mathcal{G}}\|_{i2}^2 < 1$.

Proof: The proofs are the same as those of the sufficiency parts of Theorems 3.1 and 3.4. That is, if the first Riccati equation (3.34) has a solution X_k, such that Z_k is positive definite, then X_k is necessarilly positive semidefinite. Moreover, the optimization problem (3.40) exists, and if the second Riccati equation (3.45) has a solution, then $\overline{J}_2 - x_0 X_0 x_0 < 0$. Finally, if the third condition is satisfied, namely $\frac{\gamma^2}{2} \overline{X}_0 \geq X_0$, then

$$\overline{J}_2 - \frac{\gamma^2}{2} x_0' \overline{X}_0 x_0 < 0$$

This would imply that $\|\overline{\mathcal{G}}\|_{i2}^2 < 1$.   ◇

The robust a posteriori estimator that gives the filtered estimate has the same first Riccati equation as the one-step predictor, namely, Eq.(3.34). The second Riccati equation is Eq.(3.45). The dynamics and gain equations are the same as those of Eqs.(3.17a, 3.17b), but with the new overbarred matrices substituted, as is done for the one-step predictor.

3.3.3   Robust Estimator Equations and Discussion

The equations for the robust filter derived in this section are now summarized,

$$\hat{x}_{k+1} = (\overline{A}_k - K_k\overline{C}_k)\hat{x}_k + K_k y_k \tag{3.46}$$

$$K_k = \left(\overline{A}_k H_k \overline{C}'_k + \overline{B}_k \overline{D}'_k\right)$$

$$\left(\overline{C}_k H_k \overline{C}'_k + \overline{D}_k \overline{D}'_k\right)^{-1} \tag{3.47}$$

where

$$\overline{A}_k = A_k + \gamma^{-2}B_k Z_k^{-1}F'_k \tag{3.48}$$

$$\overline{B}_k = B_k Z_k^{-1/2} \tag{3.49}$$

$$\overline{C}_k = C_k + \gamma^{-2}D_k Z_k^{-1}F'_k \tag{3.50}$$

$$\overline{D}_k = D_k Z_k^{-1/2} \tag{3.51}$$

$$F_k = S'_k T_k + A'_k X_{k+1} B_k \tag{3.52}$$

$$H_k = \left(P_k^{-1} - \gamma^{-2}M'_k M_k\right)^{-1} \tag{3.53}$$

$$Z_k = I - \gamma^{-2}(T'_k T_k + B'_k X_{k+1} B_k) \tag{3.54}$$

and the matrices X_k and P_k are obtained from the following two Riccati equations:

$$X_k = A'_k X_{k+1} A_k + S'_k S_k + \gamma^{-2}F_k Z_k^{-1}F'_k$$

$$X_N = 0 \tag{3.55}$$

$$P_{k+1} = (\overline{A}_k - K_k\overline{C}_k)H_k(\overline{A}_k - K_k\overline{C}_k)'$$

$$+(\overline{B}_k - K_k\overline{D}_k)(\overline{B}_k - K_k\overline{D}_k)'$$

$$P_0 = \text{Weight on initial error} \tag{3.56}$$

In the above, γ must be less than one, and such that $Z_k > 0$, $H_k > 0$, and $X_0 < \gamma^2\overline{X}_0$.

The solution to this problem is an extension of both the game theoretic or H_∞ optimal estimator and the Kalman filter for nominal systems. If there are no model perturbations, then $S_k = 0, T_k = 0$ in Eq.(3.29), so that the Riccati equation (3.55) for X_k is superfluous, i.e. $X_k = 0$. In that case, $\overline{X}_0 = 0$ necessarily. The estimator is reduced to solving one Riccati equation based on the nominal plant dynamics. The Riccati equation (3.13a) and the gain (3.15) are then the same as those derived in the last section.

The robust estimator can therefore be viewed as a further extension to the Kalman filter that permits the designer to trade off nominal performance in the minimum error variance sense to provide robustness to disturbance *and* plant modeling errors.

3.4 Extension to Steady State

In this section, we extend our results to stable linear time-invariant systems on an infinite-time horizon. In Section 2.2.4 we mentioned that in the frequency domain, the induced 2-norm is identified with the H_∞ norm. We first prove this result. Next, we discuss the extension of the previous section's results to steady state.

Theorem 3.6 *Let a stable linear system have transfer matrix $G_{er}(z)$, and let $\mathcal{G}$ denote the linear mapping from the square summable input sequence to the square summable output sequence. Then the induced 2-norm of $\mathcal{G}$ coincides with the H_∞ norm of $G_{er}(z)$. Specifically,*

$$\sup_{r \neq 0} \frac{\|e\|_2}{\|r\|_2} \equiv \|\mathcal{G}\|_{i2} = \|G_{er}(z)\|_\infty \equiv \sup_\omega \sigma_{max}\left(G_{er}(e^{j\omega})\right) \qquad (3.57)$$

Proof: Denote by r the input signal and $e = \mathcal{G}r$ the output signal. Let $\check{r}$ and $\check{e}$ be the Fourier transform of the sequences r_k and e_k, $k = 0, 1, \ldots$, respectively. Then

$$
\begin{aligned}
\sum_{k=1}^\infty e_k' e_k &= \frac{1}{2\pi} \int_{-\pi}^{+\pi} \check{e}'\left(e^{j\omega}\right) \check{e}\left(e^{-j\omega}\right) d\omega \\
&= \frac{1}{2\pi} \int_{-\pi}^{+\pi} \left(G_{er}(e^{j\omega})\check{r}(e^{j\omega})\right)' \left(G_{er}(e^{-j\omega})\check{r}(e^{-j\omega})\right) d\omega \\
&\leq \frac{1}{2\pi} \int_{-\pi}^{+\pi} \sigma_{max}^2 \left(G_{er}(e^{j\omega})\right) \check{r}'(e^{j\omega})\check{r}(e^{-j\omega}) d\omega \\
&\leq \frac{1}{2\pi} \sup_\omega \sigma_{max}^2 \left(G_{er}(e^{j\omega})\right) \int_{-\pi}^{+\pi} \check{r}'(e^{j\omega})\check{r}(e^{-j\omega}) d\omega \\
&= \|G_{er}(z)\|_\infty^2 \sum_{k=0}^\infty r_k r_k'
\end{aligned}
$$

The first equality follows from Parseval's identity. The second equality is simply obtained by substituting for $\check{e}(e^{j\omega})$. The first inequality follows from the properties of singular values, while the second inequality is obvious.

The last equality is obtained again using Parseval's identity. We see that $\|\mathcal{G}\|_{i2} \leq \|G_{er}(z)\|_\infty$.

Conversely, let us suppose that $\gamma < \|G_{er}\|_\infty$. This implies that there exists some frequency Ω for which $\sigma_{max}\left(G_{er}(e^{j\Omega})\right) > \gamma$. By continuity, there must also exist some $\eta > 0$ such that $\sigma_{max}\left(G_{er}(e^{j\omega})\right) > \gamma$ for all ω in the interval $[-\Omega - \eta, -\Omega + \eta]$ as well as in the interval $[\Omega - \eta, \Omega + \eta]$. Choose a disturbance that is zero outside this frequency range, and such that for each frequency inside the range, it coincides with the eigenvector corresponding to the largest eigenvalue of $G'_{er}(e^{-j\omega})G_{er}(e^{j\omega})$ for all $\omega \in [-\Omega - \eta, -\Omega + \eta]$. Then the corresponding output e is given by

$$
\begin{aligned}
\|e\|_2^2 &= \frac{1}{2\pi}\int_{-\pi}^{+\pi}\breve{e}'(e^{-j\omega})\breve{e}(e^{-j\omega})d\omega \\
&= \frac{1}{2\pi}\left(\int_{-\Omega-\eta}^{-\Omega+\eta}\breve{e}'(e^{j\omega})\breve{e}(e^{-j\omega})d\omega + \int_{\Omega-\eta}^{\Omega+\eta}\breve{e}'(e^{j\omega})\breve{e}(e^{-j\omega})d\omega\right) \\
&\geq \frac{1}{2\pi}\left(\int_{-\Omega-\eta}^{-\Omega+\eta}\gamma^2\breve{r}'(e^{j\omega})\breve{r}(e^{-j\omega})d\omega + \int_{\Omega-\eta}^{\Omega+\eta}\gamma^2\breve{r}'(e^{j\omega})\breve{r}(e^{-j\omega})d\omega\right) \\
&= \gamma^2\|r\|_2^2
\end{aligned}
$$

Hence $\|\mathcal{G}\|_{i2} \geq \|G_{er}(z)\|_\infty$. $\diamond$

The theorem shows that in steady state, the robust estimation problem formulation can be expressed in terms of bounds on the H_∞ norm of the transfer function between the disturbance r and the error e. Specifically, the performance criterion for steady state is similar to that of Eq.(3.1), or Eq.(3.27), except that the initial condition does not appear. We therefore have

$$
\begin{aligned}
\mathcal{G}_{i2} &\equiv \sup_{r\neq 0}\frac{\|e\|}{\|r\|} \\
&\equiv \|G_{er}\|_\infty \\
&= \sup_w \sigma_{max}\left(G_{er}(e^{j\omega})\right) \\
&< 1 \\
\forall\Delta \ni \|\Delta\| &\equiv \sup_{\epsilon\neq 0}\frac{\|\eta\|}{\|\epsilon\|} \\
&\equiv \|\Delta\|_\infty \\
&< 1
\end{aligned} \tag{3.58}
$$

For the sake of comparison, recall that the Kalman filter minimizes the 2-norm of the estimation error squared. In steady state, this becomes the H_2 norm of the transfer function G_{er} squared, or

$$\|G_{er}\|_2^2 \equiv \frac{1}{2\pi} \int_{-\infty}^{+\infty} \left(\sum_{i=1}^{N} \sigma_i^2 \left(G_{er}(e^{j\omega}) \right) \right) d\omega$$

When the plants under consideration are linear time invariant, so that $A_k = A$, $B_k = B$, $C_k = C$, $D_k = D$, $S_k = S$, $T_k = T$ and $M_k = M$ for all $k = 0, 1, ...$ then the estimation problem can be defined as follows: Does an estimator that achieves an H_∞ bound on the mapping between the disturbance and the error exist? If so, does that estimator yield stabilizing error dynamics?

For the first question above to be meaningfull, the answer to the second must be positive since any estimator that yields unstable error dynamics is useless regardless of its other properties. Now, if $\hat{x}_{k+1} = A_e \hat{x}_k + K_e y_k$ is the estimator state equation, then the augmented state and error dynamics are given by

$$\begin{bmatrix} x_{k+1} \\ \tilde{x}_{k+1} \end{bmatrix} = \begin{bmatrix} A & 0 \\ A - (A_e + K_e C) & A_e \end{bmatrix} \begin{bmatrix} x_k \\ \tilde{x}_k \end{bmatrix}$$

$$+ \begin{bmatrix} B \\ B - K_e D \end{bmatrix} r_k \qquad (3.59)$$

If the plant is stable, then any stable estimator produces stable error dynamics even if the plant model is uncertain. If the plant is unstable but stabilizable, detectable, and its model is known, then setting $A_e = A - K_e C$ and choosing K_e so that A_e is stable will produce stable error dynamics. If on the other hand the plant is both uncertain and unstable, then as discussed in Section 2.2.3, no estimator will stabilize the error dynamics. Thus, the steady state robust estimation problem is meaningfull for stable plants only.

We now turn to the first question asked above, specifically the existence of an estimator that achieves an H_∞ bound on G_{er}. The answer to this question depends on the existence of solution to the two steady state Riccati equations:

$$X = A'XA + S'S + \gamma^{-2}FZ^{-1}F' \qquad (3.60)$$

$$P = \left(\overline{A} - K\overline{C} \right) H \left(\overline{A} - K\overline{C} \right)' + \left(\overline{B} - K\overline{D} \right) \left(\overline{B} - K\overline{D} \right)' \qquad (3.61)$$

where

$$F \equiv S'T + A'XB \qquad (3.62)$$

$$Z \equiv I - \gamma^{-2}(T'T + B'XB) \tag{3.63}$$

$$H \equiv \left(P^{-1} - \gamma^{-2}M'M\right)^{-1} \tag{3.64}$$

$$K = (\overline{A}H\overline{C}' + \overline{BD'})(\overline{C}H\overline{C}' + \overline{DD'})^{-1} \tag{3.65}$$

and the overbarred matrices are the same as those given by Eqs.(3.48-3.51), with subscripts k omitted. It is possible to show that if the plants under consideration are all stable and, in addition, they satisfy the conditions of the previous section (such as invertibility of the matrices A and $CHC' + DD'$), then the desired robust estimator exists if 1) Eq.(3.60) admits a solution such that X is positive semidefinite and Z is positive definite, and 2) Eq.(3.61) admits a solution such that P and H are positive definite. We emphasize that these conditions are sufficient, but not necessary. The steady-state estimator is then given by

$$\hat{x}_{k+1} = (\overline{A} - K\overline{C})\hat{x}_k + Ky_k \tag{3.66}$$

In the absence of model uncertainty, then only the Riccati equation in P matters. Furthermore, with the plant model known, the plant needs not be stable. Instead, it is possible to show that if (A, C) is detectable, (A, B) is stabilizable, and the matrix

$$\begin{bmatrix} A - e^{j\omega}I & B \\ C & D \end{bmatrix}$$

is of full row rank for all $\omega \in [0, 2\pi)$, then an estimator exists if the steady state Riccati equation given below admits a solution such that P and H are positive definite:

$$P = (A - KC)H(A - KC)' + (B - KD)(B - KD)' \tag{3.67}$$

$$K = (AHC' + BD')(CHC' + DD')^{-1} \tag{3.68}$$

$$H = \left(P^{-1} - \gamma^{-2}M'M\right)^{-1} \tag{3.69}$$

The state estimator equation is then given by

$$\hat{x}_{k+1} = (A - KC)\hat{x}_k + Ky_k \tag{3.70}$$

Note that when γ is at its minimum value, then we have the H_∞ optimal estimator. The results stated above are similar in nature to those of the steady state discrete time control problem, which is discussed in [104].

3.5 Robust Fixed-Interval Smoothing

In the previous sections of this chapter, the state estimates we sought are either strictly causal, or causal. Specifically, $\hat{x}_k$ is a function of $y_0, \ldots, y_{k-1}$, and $\hat{x}_{k|k}$ is a function of $y_0, \ldots, y_k$. In some situations, it may be beneficial to obtain estimates that are a function of past, present, and future measurements. These are called smoothed estimates. There are three types of smoothed estimates: fixed lag, fixed point, and fixed interval. In Section A.8 we describe the three problems and show that the first two can be recast as filtering problems. The results of the previous sections can therefore be used to obtain the fixed-point and fixed-lag game theoretic smoothed estimates. We shall therefore concentrate in this section on game theoretic fixed-interval smoothing. Perhaps the main point, or the surprise of this section is that for fixed-interval smoothing, the game theoretic and linear minimum error variance (Kalman) estimates are one and the same!

In fixed-interval smoothing, our objective is to obtain an estimate of the state trajectory over an entire time interval as a function of measurements for the entire interval. Specifically, if $Y \equiv [y_0, \ldots, y_k, \ldots, y_N]$, then our objective is to find estimate of the trajectory, or $\hat{x}(Y) \equiv [\hat{x}_0(Y), \ldots, \hat{x}_k(Y), \ldots, \hat{x}_N(Y)]$. We first derive a game theoretic fixed-time interval smoother based on a nominal plant model, and then show that the results obtained can be extended to give a smoother that is robust to plant model uncertainties as well.

In deriving the filtering equations of Sections 3.2 and 3.3, we have assumed that the process and measurement noise are correlated, or equivalently that $B_k D'_k \neq 0$ for at least some k. We shall assume here that they are uncorrelated, so that $B_k D'_k \equiv 0$ for all k. This will simplify the derivation. We will show later, using a transformation of variables, as is done in [68], that there is no loss of generality resulting from this assumption. One can therefore take w_k to be the process noise and v_k the measurement noise at time k.

The smoothing problem can be formulated as

$$\min_{x^s(Y)} \max_{v, w, x_0} \sum_{k=1}^{N} \frac{1}{2} \|x_k - x_k^s(Y)\|_{M'_k M_k}^2 \tag{3.71}$$

subject to

$$\sum_{k=0}^{N-1} \|w_k\|_{Q_k^{-1}}^2 + \|v_k\|_{V_k^{-1}}^2 + \|x_0 - \hat{x}_0\|_{P_0^{-1}}^2 = 1 \tag{3.72}$$

and

$$x_{k+1} = A_k x_k + B_k w_k \tag{3.73}$$
$$y_k = C_k x_k + D_k v_k \tag{3.74}$$

In Eq.(3.72), the invertible matrices Q_k and V_k are norm weights. Substituting for v_k from Eq.(3.74), the augmented cost function is, for $k = 0, ..., N-1$

$$
\begin{aligned}
\tilde{J}_s = & \sum_{k=1}^{N} \|x_k - x_k^s(Y)\|_{M_k'M_k}^2 \\
& - \frac{\gamma^2}{2} \left(\sum_{k=0}^{N-1} \|w_k\|_{Q_k^{-1}}^2 + \|y_k - C_k x_k\|_{\overline{V}_k^{-1}}^2 \right) \\
& + \sum_{k=0}^{N-1} \gamma^2 \lambda_{k+1}^s (x_{k+1} - A_k x_k - B_k w_k) \\
& - \frac{\gamma^2}{2} \|x_0 - \hat{x}_0\|_{P_0^{-1}}^2
\end{aligned}
\tag{3.75}
$$

where γ^2 and the $\gamma^2 \lambda_k$'s are the Lagrange multipliers, and $\overline{V}_k = \left(D_k V_k^{-1} D_k' \right)^{-1}$. The minmax or game theoretic smoother must account for the worst possible process noise w_k and initial condition x_0. The corresponding measurement noise is accounted for through the term $\|y_k - C_k x_k\|_{\overline{V}_k^{-1}}^2$ in the objective function.

In terms of the game formulation, therefore, the problem reduces to

$$\min_{x^s(Y)} \max_{w, x_0} \tilde{J}_s \tag{3.76}$$

Taking the appropriate first variation, the necessary conditions are

$$x_{k+1} = A_k x_k + B_k w_k \tag{3.77a}$$
$$w_k = Q_k B_k' \lambda_{k+1}^s \tag{3.77b}$$
$$-\lambda_k^s = \left(-\gamma^{-2} M_k' M_k + C_k' \overline{V}_k^{-1} C_k \right) x_k - A_k' \lambda_{k+1}$$
$$\qquad + \gamma^{-2} M_k' M_k x_k^s + C_k' \overline{V}_k^{-1} y_k \tag{3.77c}$$
$$x_0 = \hat{x}_0 + P_0 \lambda_0 \tag{3.77d}$$
$$\lambda_{N+1}^s \quad \text{free} \tag{3.77e}$$

and, from the variation with respect to $x_k^s(Y)$,

$$x_k = x_k^s(Y) \tag{3.78}$$

The last equation indicates that the smoothed estimate follows the state trajectory that is generated by the worst possible process noise and initial state. Eqs.(3.77a-3.78) give the Hamiltonian

$$
\begin{bmatrix} x_{k+1}^s \\ -\lambda_k^s \end{bmatrix} = \begin{bmatrix} A_k & B_k Q_k B_k' \\ C_k' \overline{V}_k^{-1} C_k & -A_k' \end{bmatrix} \begin{bmatrix} x_k^s \\ \lambda_{k+1}^s \end{bmatrix}
$$
$$
+ \begin{bmatrix} 0 \\ -C_k' \overline{V}_k^{-1} \end{bmatrix} y_k \tag{3.79}
$$

with boundary conditions

$$
x_0^s = \hat{x}_0 + P_0 \lambda_0 \tag{3.80}
$$
$$
\lambda_{N+1} \quad \text{free} \tag{3.81}
$$

This is exactly the same Hamiltonian obtained when solving the minimum error variance smoothing problem, as is shown in Section A.8.1. The game theoretic and the Kalman fixed-interval smoothers are therefore identical. A possible explanation of this equivalence should lie in the fact that the measurement history for the entire interval is available. This measurement history, Y, is used by both estimators to obtain the input disturbance sample path. In the case of the game theoretic fixed-interval smoothing problem, the worst possible sample path the disturbance can take, given Y, is no worse than the least squares error estimate of the disturbance, given Y. In other words, the availability of Y possibly excludes the worst disturbance sample paths that could be obtained were the measurements not available, as is the case with the filtering problem.

Note that the solution to the game theoretic fixed-lag smoothing problem is not the same as the Kalman smoother [45]. The same is true for the fixed-point smoothing problem. As mentioned before, both these problems can be seen as filtering problem, as is shown in Section A.8.2.

To obtain a recursive formulation for the smoother, the Hamiltonian can be triangularized using the transformation

$$
\hat{x}_k = x_k^s - P_k \lambda_k^s \tag{3.82}
$$

where $\hat{x}_k$ is the one-step prediction estimate, and P_k is the Kalman filter's prediction error covariance at time k. This gives

$$
\begin{bmatrix} \hat{x}_{k+1} \\ -\lambda_k^s \end{bmatrix} = \begin{bmatrix} A_k - K_k C_k & 0 \\ \breve{F}_k C_k' \overline{V}_k^{-1} C_k & -\breve{F}_k A_k' \end{bmatrix} \begin{bmatrix} \hat{x}_k \\ \lambda_{k+1}^s \end{bmatrix}
$$

$$+ \begin{bmatrix} K_k \\ -\check{F}_k C_k' \overline{V}_k^{-1} \end{bmatrix} y_k \tag{3.83}$$

where

$$K_k = A_k P_k C_k' \left(C_k P_k C_k' + \overline{V}_k \right)^{-1} \tag{3.84}$$

$$\begin{aligned} P_{k+1} &= (A_k - K_k C_k)\, P_k \, (A_k - K_k C_k)' \\ &\quad + \left(\Gamma_k - K_k \overline{V}_k^{1/2} \right) P_k \left(\Gamma_k - K_k \overline{V}_k^{1/2} \right)' \end{aligned} \tag{3.85}$$

$$\check{F}_k = \left(-I + C_k' \overline{V}_k^{-1} C_k P_k \right)^{-1} \tag{3.86}$$

Here $\Gamma_k = B_k Q_k^{1/2}$. The initial condition is

$$\hat{x}_0 = \text{given} \tag{3.87}$$

$$\lambda_{N+1}^s \quad \text{free} \tag{3.88}$$

Note that we have the Kalman filter at the top row of the triangularized Hamiltonian. With the forward sweep, we have the one-step prediction estimates $\hat{x}_1, \ldots, \hat{x}_N$. A backward sweep using the bottom row of Eq.(3.83) gives the values of λ_k^s. To start the backward sweep, we assume that for the fictitious step from N to $N+1$, we have $A_N = I$ and $B_N = 0$. As a result $\hat{x}_{N|N} = \hat{x}_{N+1}$. We also have $\hat{x}_{N|N} = \hat{x}_N^s$. With the last measurement taken at time N, it then follows that $\hat{x}_{N+1}^s = \hat{x}_N^s = \hat{x}_{N+1}$. The backward sweep can therefore be started with the guess $\lambda_{N+1}^s = 0$. The smoothed estimates x_k^s for all k follow from Eq.(3.82). Other forms of the Hamiltonian equation are shown in Section A.8.

We now show how to transform a system where the process and measurement noise are correlated into one with the form of Eq.(3.73). Consider the dynamic system

$$\begin{aligned} x_{k+1} &= A_k x_k + B_k r_k \\ y_k &= C_k x_k + D_k r_k \end{aligned}$$

Multiplying the second equation above by $\check{L}_k \equiv B_k D_k' (D_k D_k')^{-\frac{1}{2}}$, and subtracting it from the first gives a new state space equation. Defining $\check{y}_k \equiv (D_k D_k')^{-\frac{1}{2}} y_k$ gives a new set of equations

$$\begin{aligned} x_{k+1} &= \check{A}_k x_k + \check{B}_k r_k + \check{L}_k y_k \\ \check{y}_k &= \check{C}_k x_k + \check{D}_k r_k \end{aligned}$$

where

$$
\begin{aligned}
\check{A}_k &\equiv A_k - \check{L}_k C_k \\
\check{B}_k &\equiv B_k - \check{L}_k D_k \\
\check{C}_k &\equiv (D_k D_k')^{-\frac{1}{2}} C_k \\
\check{D}_k &\equiv (D_k D_k')^{-\frac{1}{2}} D_k
\end{aligned}
$$

and $\check{B}_k \check{D}_k' = 0$. We now take B_k and D_k to be identity matrices. Now, with $w_k = \check{B}_k r_k$, $v_k = \check{D}_k r_k$, $Q_k = \check{B}_k \check{B}_k'$, and $V_k = \check{D}_k \check{D}_k'$, we have the form of Eqs.(3.73,3.74), except for a known input term, which can be tagged along. We note that $\check{L}_k$ can be any matrix that satisfies $\left(B_k - \check{L}_k D_k \right) (D_k D_k')^{-1} D_k = 0$.

The results obtained above can be generalized to the case where the plant dynamics are uncertain. In that case, the robust performance criterion is the same as that of Eq.(3.26), except that the error at each time k is defined as $e_k = M_k (x_k - x_k^s(Y))$. This leads to the optimization problem of Eq.(3.32), subject to the dynamic constraints of Eq.(3.29), but with $x^s(Y)$ replacing $\hat{x}$. Specifically, the robust smoothing problem leads to the following objective function

$$
\min_{\hat{x}(Y)} \max_{d, x_0} \left\{ \sum_{k=1}^{N} \|x_k - x^s(Y)\|_{M_k' M_k}^2 + \|S_k x_k + T_k d_k\|^2 \right.
$$
$$
\left. -\gamma^2 \sum_{k=0}^{N-1} \|d_k\|^2 + \|x_0\|_{X_0}^2 + \|x_0 - \hat{x}_0\|_{P_0^{-1}}^2 \right\}
$$

The first Riccati equation (3.34), together with the transformations of Eqs.(3.39a-3.39d), can then be used to reduce the problem to the one solved in this section.

3.6 Numerical Examples

In this section, we present two examples. The first example discusses a robust estimator design for the two-state system of Section 2.2.3. The second example is an application of robust filtering to navigation, or more precisely, attitude determination. In the second example, we also demonstrate that ad hoc methods commonly used to robustify the Kalman filter do not give the best results.

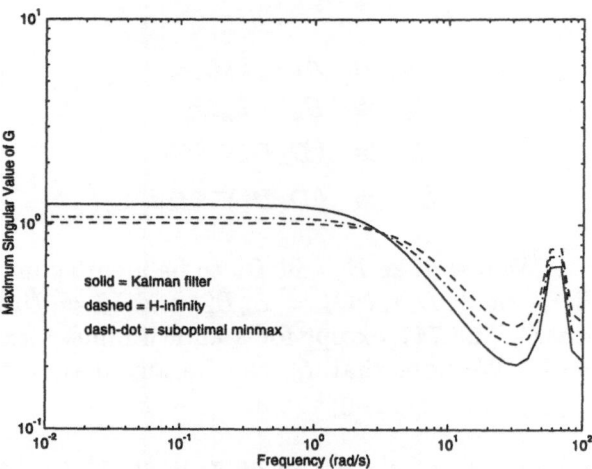

Figure 3.3: Frequency response from noise to estimation error for a nominal plant.

3.6.1   A Two-State System

The numerical example of Section 2.2.3 is now revisited in order to compare the Kalman filter with the estimators derived in this chapter. Consider first the nominal plant. Estimators with robustness to noise model uncertainty only are used to estimate the state of these plants at steady state. Figure 3.3 shows the maximum singular value of the mapping from the noise to the estimation error for the Kalman filter, with $\gamma = \infty$; the H_∞ optimal estimator, with $\gamma = \gamma_{min}$; and a third estimator, with an intermediate value of γ. The H_∞ estimator minimizes the peak gain over frequency, which can be seen in the figure. Reducing the peak gain, however, comes at the expense of less rejection of the higher frequency noise. The intermediate curve shows the result of choosing γ above the optimal level and below infinity (the Kalman filter).

The same simple academic example of Section 2.2.3 is used to compare the performance of the robust estimator to the Kalman filter and H_∞ optimal estimators for both the nominal and perturbed dynamic models. Figure 3.4 shows the maximum singular value response of these three estimators with nominal plant dynamics. Figure 3.5 shows the response with a large perturbation in the plant dynamics.

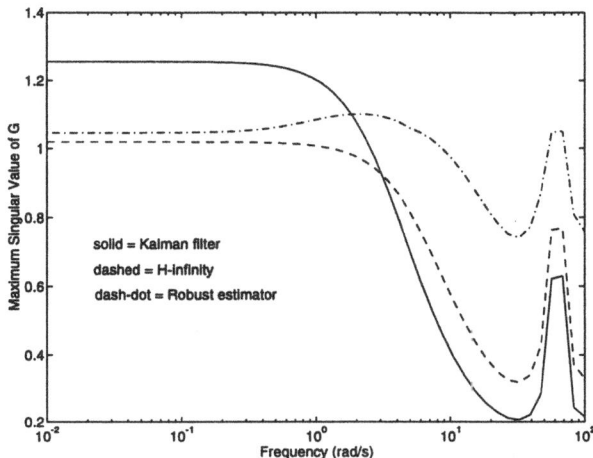

Figure 3.4: Frequency response for nominal plant.

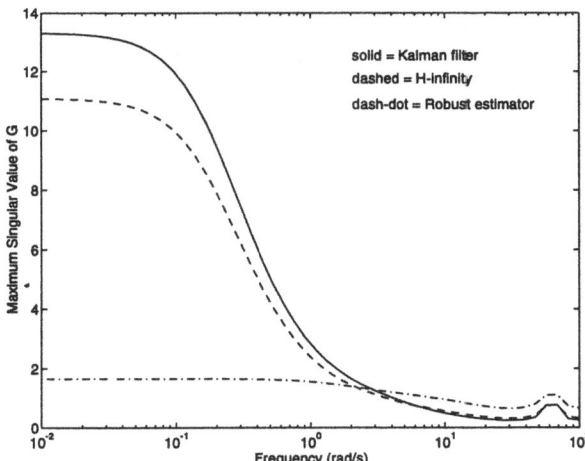

Figure 3.5: Frequency response for perturbed plant.

Note the large degradation in noise rejection with both the Kalman filter and H_∞ optimal estimator with the perturbed plant model. The noise rejection performance with the robust estimator, however, is relatively insensitive to the plant perturbation. The expense is less rejection of noise, particularly at higher frequencies, for the nominal plant. For practical applications, however, insensitivity to plant modeling errors is well worth the modest degradation in nominal performance.

3.6.2   Attitude Determination

Navigation has been one of the most successful applications of the Kalman Filter since its birth [83]. The purpose of a navigation filter is to provide information on the moving vehicle's position, velocity, attitude, and attitude rate. A widely used approach to navigation filter design is the following: an inertial navigation system (INS), with its accelerometers and gyros, provides the vehicle with velocity, position, attitude rate and attitude information. The velocity is obtained by integrating the output of the accelerometer once, and the position is deduced by integrating it twice. Likewise, the gyro provides attitude rate and, after integration, attitude. Unfortunately, accelerometer and gyro biases, misalignment, and other outside factors degrade the quality of the information supplied by the INS. For this reason, external navigation aids, such as Global Positioning System (GPS) and radio navigation information, are used to compensate for the bias, misalignment, etc.. External aids usually provide at least the vehicle's position, but lately, GPS has been used to provide velocity and attitude information as well. The idea is to use IMU's for high frequency maneuvers, while external aids are used to correct for the biases and misalignments in the long term.

In the approach briefly described above, the navigation filter does not make use of the vehicle dynamic model as a source of information on the vehicle's motion. As such, it is conceptually autonomous from the vehicle on which it operates. This autonomy is favored primarily because the instruments can be tested at will in the laboratory. The test results are then utilized to accurately model the instrument errors, and a Kalman filter design based on these error models provides estimates of the biases and other errors online. The error estimates are in turn used to adjust the instruments' outputs. This method is preferred because vehicle testing is expensive, and obtaining accurate dynamic models may not be possible. Navigation filters based on the H_∞ criterion are discussed in [39].

There are, however, situations when the use of the vehicle's dynamics may be beneficial, if not necessary. This is the case, for instance, when outside navigation is interrupted frequently, or for too long a period. Model based Kalman filters for attitude determination that utilize plant dynamics are discussed in [69], and [84].

In this example, we illustrate the use of robust estimators for model based attitude determination in situations like the shuttle reentry. We consider the space shuttle Orbiter's lateral dynamics for bank and sideslip. A linear model of the rigid body reentry rotational dynamics is used as derived by Zacharias [124]. While originally given in continuous time, the dynamics are discretized for use with the Orbiter's digital computing system. The application is described in more detail in Chapter 6 and Appendix D.

Navigation in the reentry phase is difficult for several reasons. First, GPS and other outside sources of information are not available during reentry. Moreover, the rapid variation in Mach number produces large changes in the vehicle aerodynamic properties. These properties are uncertain, which makes the vehicle model inaccurate. Even if these properties are known, rapid gain scheduling of a time-varying Kalman filter may not be desirable, or even possible.

Estimation results were obtained by Agustin [2], and are shown in Table 3.1 and Figure 3.6. The nominal model is for a Mach number of $M = 7.5$, and an angle of attack of $\alpha = 35$ degrees. The perturbed plant is for $M = 8.8$ and $\alpha = 38$ degrees. At each time step, three measurements are available: The bank and sideslip angle rates, and the change in angle of attack (The second measurement, however, is not needed).

Two Kalman filters and a robust filter are compared. All filters are applied to both plants. The first Kalman filter is designed for the nominal plant model. The second Kalman filter design is based on the nominal plant as well, except that the process noise covariance is increased by a factor of 10,000 (or B_k is replaced by $100 \times B_k$). This "overdesigned" [58] filter gives more weight to the measurement data in order to protect against plant uncertainty. The robust filter was designed for an entire family of plants that includes both the nominal and the perturbed models. This is described in [2].

Table 3.1 compares the mean squared errors for the sideslip angle, the sideslip angle rate, and the bank angle rate obtained from simulation results for the three filters. The bank angle error was nearly zero

		Squared Error	Bank Rate	Sideslip Angle	Sideslip Rate
KF	Nominal Plant	0.09	0.61	0.04
	Perturbed Plant	5.52	15.21	21.93
OKF	Nominal Plant	0.37	0.52	0.08
	Perturbed Plant	0.37	3.98	1.00
ROB	Nominal Plant	0.14	0.62	0.22
	Perturbed Plant	0.15	0.97	0.24

Table 3.1: Comparison of mean squared estimation error among the Kalman filter (KF), the overdesigned Kalman filter (OKF), and the Robust filter (RF). Units are $(\deg)^2$ and $(\deg/\sec)^2$.

for all filters. It is clear that although the robust performance of the overdesigned Kalman filter is superior to that of the nominal Kalman filter, it is still much poorer than that of the robust filter. By contrast, the nominal performance of the robust and overdesigned Kalman filters is comparable. A comparison between the robust and nominally optimal Kalman filter also indicates that overall, the robust filter is preferable. Figure 3.6 shows simulation results for the robust filter. It is clear that the performance is very similar for both plants. The filters were also tried on fictitious intermediate plants, and the results were similar in nature.

We now discuss how the robust filter is designed. The perturbation was represented as a parametric uncertainty. Let the matrices A_n, B_n, C_n, and D_n represent the nominal plant. Similarly, the matrices A_p, B_p, C_p, and D_p represent the perturbed plant. Define

$$\Delta A(i,j) = .01 \times |A_n(i,j) - A_p(i,j)|$$
$$\Delta B(i,j) = .05 \times |A_n(i,j) - A_p(i,j)|$$
$$\Delta C = .0001 \times C_n$$
$$\Delta D = .005 \times D_n$$

These matrices are then used to build Q, R, S, and T matrices, as discussed in the example of Section 3.3.1. Since we seek a steady state filter for a time-invariant plant, we march in time the first Riccati equation (3.55) until it reaches steady-state, using a value of $\gamma = .083$. Next, the matrices $\overline{A}, \overline{B}, \overline{C}$, and $\overline{D}$ are obtained as summarized in Section 3.3.3, and we use them to design a Kalman filter. This means that for the second Riccati equation, γ is set to infinity.

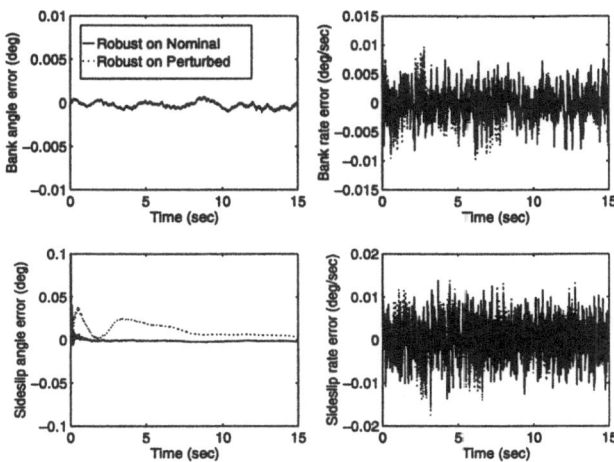

Figure 3.6: Estimation errors for robust filter on a nominal and on a perturbed plant

The reader will no doubt notice that we parted from the theory. First, the size of the uncertainty we used for the design is miniscule relative to the actual size. Second, we define parametric perturbation terms differently. Third, we do not use the same value of γ in the two Riccati equations. These measures gave the best results.

The reason for this departure has to do with the very nature of H_∞ or game theoretic optimization, which is concerned with the worst case possibility. Thus, in designing a filter, we focus on the most detrimental perturbation. Worrying about the worst possibility may be appropriate in control, where plant stability is an issue. In the case of estimation, however, the objective is robust performance, and guaranteeing the worst case performance may not make sense if it severely degrades average performance. Even in control design, worrying about the highly unlikely worst case situation may degrade robust performance in most situations. Moreover, we used a Kalman filter for the second step in the design because the white noise assumption was appropriate.

All this means that we no longer have the guarantees provided by the small gain theorem. Nevertheless, it is clear that the robust estimator, which is based on the solution of two Riccati equations, is preferable to the overdesigned Kalman filter. We have therefore used the theory as a guide on how to desensitize our design to plant uncertainty.

This example suggests one approach for robust filter design. Specifically, one can first select a plant from the family of plants of interest, design for it, then iteratively, and individually, increase the value of the design perturbation matrices, and modify the parameter γ. The candidate robust filters should then be tried on many plants of interest.

3.7   Related Work

Early work on H_∞ estimation was done by Grimble, Ho, and El-saayed [43], [44]. Other work includes that of Speyer, Chih-hai, and Banavar [99], Habibi, Sayed, and Kailath [49], [50], and Yaesh and Shaked [120], [121].

The authors in [99] solved the risk sensitive optimal estimation problem (see also [111], [112] for the risk sensitive optimal control problem). As will be seen in the next chapter, this problem is equivalent to the game theoretic estimation problem for the case where the plant model dynamics are assumed known. Habibi, Sayed, and Kailath [49], [50] used Krein space methods to obtain their solution. Though the results are similar, the approach is geometric. The work in [121] is based on a frequency-domain approach. In [120], the authors use a game theoretic approach in which the adversary uses a fictitious measurement signal to select the input disturbance. As such, their estimator is of a different nature than that of Section 3.2.

Discrete-time estimators that are robust to plant perturbations, as in Section 3.3, were developed by Xie, De Souza, and Fu [119]. They are valid only for linear time-invariant systems at steady state and a restricted class of model uncertainties. The authors in [119] use a spectral factorization approach. Finally, Grimble [45] derives results for the fixed-lag H_∞ smoother, assuming accurate knowledge of the plant model.

CHAPTER 4

STOCHASTIC INTERPRETATION OF ROBUST ESTIMATION: RISK SENSITIVITY

4.1 Introduction

The game theoretic robust estimation problem defined in Chapter 3 is a deterministic one. That is, all the input disturbances have deterministic bounds. Nevertheless, as was mentioned in Chapters 2 and 3, this class of problems is closely related to a class of stochastic estimation problems, namely, risk sensitive optimal estimation [99], [111], [112]. In risk sensitive estimation, the noise is assumed to be white, Gaussian, and the cost criterion is the expected value of the *exponential* of a weighted sum of the squared state estimation errors.

Our objective in this chapter is to demonstrate the relationship between risk sensitive and game theoretic estimation. The relationship between the two classes of problems adds insight by providing a stochastic interpretation to game theoretic estimation (and control) problems. A complete treatment of risk sensitive control theory is found in [111], [112].

The relationship between risk sensitive and game theoretic estimation for systems with known plant dynamics is the subject of the next section. The extension of the stochastic interpretation to systems with uncertain plant dynamics is found in Section 4.3. A numerical example is given in Section 4.4, and a short summary in Section 4.5.

4.2   The Risk Sensitive Optimal Estimation Problem

The risk sensitive estimation problem is first formulated using the notation of [99]. Then, the equivalence between risk sensitive and game theoretic estimation is shown.

4.2.1   Problem Formulation

Consider the following linear, stochastic, discrete time-varying system with noisy state observations:

$$x_{k+1} = A_k x_k + w_k \qquad (4.1a)$$
$$y_k = C_k x_k + v_k \qquad (4.1b)$$

The variables x_k, w_k, y_k, and v_k are vectors of appropriate dimensions defined on the interval $k = 0, ..., N-1$. The initial state x_0 is a normally distributed random variable with mean $\hat{x}_0$ and covariance P_0. The w_k's and v_k's are zero mean, white, and jointly Gaussian processes with known joint covariance matrices S_k, where

$$S_k \equiv E\left(\begin{bmatrix} w_k \\ v_k \end{bmatrix} \begin{bmatrix} w_k' & v_k' \end{bmatrix}\right)$$
$$= \begin{bmatrix} \Theta_k & \Lambda_k \\ \Lambda_k' & R_k \end{bmatrix} \qquad (4.2)$$

The risk sensitive estimation problem is defined as

$$\min_{\hat{x}} \mathcal{L} \qquad (4.3)$$
$$\text{subject to} \quad \text{Eqs.}(4.1a, 4.1b)$$

where

$$\mathcal{L} \equiv \theta^{-1} \log E\left(e^{\theta J}\right) \qquad (4.4)$$

with

$$J \equiv \frac{1}{2} \sum_{k=1}^{N} e_k' e_k \tag{4.5}$$

$$e_k \equiv M_k(x_k - \hat{x}_k)$$

The expectation operation is with respect to the random variables x_0, w_k, and v_k. For *risk sensitive* optimization, the parameter θ is positive. In that case, a large error term has a relatively heavier weight in the objective function when compared to a small error term, because of the exponentiation in Eq.(4.4). This concern for large errors is the reason for the term "risk sensitive". Note also that this fact is a strong hint for the relationship with game theoretic estimation, and in particular H_∞ estimation, where it is assumed that the noise takes the worst possible value.

The parameter θ in Eq. (4.4) can be zero or negative as well. For θ negative, small error terms weigh more significantly than large ones. This case is called *risk seeking* or *risk prone*, meaning that the designer is optimistic, assuming that the noise will be beneficial.

Finally, as θ approaches zero from either the negative or positive side, we approach the *risk neutral* case, which is the minimum variance estimation problem. To see this, let $\zeta(\theta) = E(e^{\theta J})$, and expand $\log(\zeta(\theta))$ in a Taylor expansion around $\theta = 0$. Thus,

$$\begin{aligned} \log(\zeta(\theta)) &= \log(\zeta(0)) + \left(\frac{d}{d\theta} \log \zeta(0)\right)\theta + \left(\frac{d^2}{d\theta^2} \log \zeta(0)\right)\theta^2 \\ &= \log(\zeta(0)) + \frac{d\zeta(0)/d\theta}{\zeta(0)}\theta \\ &\quad + \frac{1}{2}\frac{d^2\zeta(0)/d\theta^2 - (d\zeta(0)/d\theta)^2(0)}{\zeta^2(0)}\theta^2 + \dots \end{aligned}$$

With $\zeta(0) = 1$, $d\zeta(0)/d\theta = E(J)$, and $d^2\zeta(0)/d\theta^2 = E(J^2)$, we have

$$\begin{aligned} \mathcal{L} &= \frac{1}{\theta}(\log \zeta(\theta)) \\ &= \frac{1}{\theta}\left(E(J)\theta + \frac{\theta^2}{2}\left(E(J^2) - E^2(J)\right) + \dots\right) \end{aligned}$$

Taking the limit as $\theta \to 0$, we obtain

$$\lim_{\theta \to 0} \mathcal{L} = E(J) \tag{4.6}$$

which says that the risk neutral case is equivalent to the problem of minimizing average performance, whose solution is the Kalman filter.

4.2.2   Equivalence to Game Theoretic Estimation

We begin by rewriting the expectation operation of the cost $\mathcal{L}$ of Eq.(4.4) in a more convenient form. Define

$$w \equiv [w_0, ..., w_{N-1}]$$
$$v \equiv [v_0, ..., v_{N-1}]$$

Then (x_0, w, v) has a joint Gaussian density function. Recalling that the w_k's and v_k's are white processes uncorrelated with x_0, we have

$$f((x_0 - \hat{x}_0), w, v) = f(x_0 - \hat{x}_0) \prod_{k=0}^{N-1} f\left(\begin{bmatrix} w_k \\ v_k \end{bmatrix} \right)$$
$$= \text{Constant} \times e^{-\Psi} \qquad (4.7)$$

where

$$\Psi \equiv \frac{1}{2}(x_0 - \hat{x}_0)' P_0^{-1}(x_0 - \hat{x}_0) + \frac{1}{2} \sum_{k=0}^{N-1} \begin{bmatrix} w_k' & v_k' \end{bmatrix} S_k^{-1} \begin{bmatrix} w_k \\ v_k \end{bmatrix}$$

The objective function of Eq.(4.4) is now:

$$\mathcal{L} \equiv \frac{1}{\theta} \log E\left(e^{\theta J}\right)$$
$$= \frac{1}{\theta} \log \int e^{\theta J} f((x_0 - \hat{x}_0), w, v) \, d(x_0 - \hat{x}_0) \, dw \, dv$$
$$= \frac{1}{\theta} \log \int e^{\theta J - \Psi} d(x_0 - \hat{x}_0) \, dw \, dv + \text{Constant} \qquad (4.8)$$

Note that the term $\theta J - \Psi$ has a quadratic form. The following theorem can be used to relate the integral in Eq.(4.8) to a maximization with respect to the integration variables (see also [99] and [112]).

Theorem 4.1 *Let $\mathcal{Z}(\alpha, \beta)$ be a quadratic function of the vector variables α, β, such that the Hessian $\mathcal{Z}_{\alpha\alpha} < 0$, then:*

$$\int \exp\left(\mathcal{Z}(\alpha, \beta)\right) d\alpha = \text{Constant} \times \exp\left(\max_\alpha \mathcal{Z}(\alpha, \beta)\right) \qquad (4.9)$$

Proof: Let α^* be the value of α at which $\mathcal{Z}$ achieves its maximum. Then, expanding $\mathcal{Z}$ in a Taylor series around α^*, we have

$$\mathcal{Z}(\alpha, \beta) = \mathcal{Z}(\alpha^*, \beta) + \frac{1}{2}(\alpha - \alpha^*)' \mathcal{Z}_{\alpha\alpha}(\alpha - \alpha^*)$$

Terms with third or higher order derivatives in the expansion vanish thanks
to the quadratic form of $\mathcal{Z}$. Now,

$$
\begin{aligned}
\int \exp\left(\mathcal{Z}(\alpha,\beta)\right) d\alpha &= \int \exp\left(\mathcal{Z}(\alpha^*,\beta) + \frac{1}{2}(\alpha-\alpha^*)'\mathcal{Z}_{\alpha\alpha}(\alpha^*,\beta)(\alpha-\alpha^*)\right) d\alpha \\
&= \exp\left(\mathcal{Z}(\alpha^*,\beta)\right) \int \exp\left(\frac{1}{2}(\alpha-\alpha^*)\mathcal{Z}_{\alpha\alpha}(\alpha^*,\beta)(\alpha-\alpha^*)\right) d\alpha \\
&= \text{Constant} \times \exp\left(\mathcal{Z}(\alpha^*,\beta)\right)
\end{aligned}
$$

$\diamond$

    The above theorem is the key relationship that makes it possible
to establish the equivalence between the risk sensitive optimal estimation problem of this chapter and the game theoretic estimation problem
of Chapter 3. The following corollary follows immediately.

Corollary 4.2

$$
\mathcal{L} = \max_{(w,v,x_0)} \left(J - \theta^{-1}\Psi\right) + \text{Constant}
$$

$\diamond$

Finally, it is clear that

$$
\min_{\hat{x}} \mathcal{L}
$$

subject to the dynamic constraints (4.1a,4.1b) is equivalent to

$$
\min_{\hat{x}} \max_{(w,v,x_0)} \left(J - \theta^{-1}\Psi\right) \tag{4.10}
$$

subject to the same dynamic constraints. The resulting problem is
nothing other than a game theoretic estimation problem. To establish
the equivalence in detail, write

$$
\begin{aligned}
S_k &= \begin{bmatrix} \Theta_k & \Lambda_k \\ \Lambda_k' & R_k \end{bmatrix} \\
&= \begin{bmatrix} B_k B_k' & B_k D_k' \\ D_k B_k' & D_k D_k' \end{bmatrix}
\end{aligned}
$$

and

$$
\begin{aligned}
w_k &= B_k r_k & \text{(4.11a)} \\
v_k &= D_k r_k & \text{(4.11b)} \\
\theta &= \gamma^{-2} & \text{(4.11c)}
\end{aligned}
$$

where r_k, $k = 0, \ldots, N - 1$ is a white noise process of appropriate dimension with unit covariance. It follows that, for $k = 0, \ldots, N - 1$,

$$
\begin{bmatrix} w'_k & v'_k \end{bmatrix} S_k^{-1} \begin{bmatrix} w_k \\ v_k \end{bmatrix} = r'_k \begin{bmatrix} B'_k & D'_k \end{bmatrix} S_k^{-1} \begin{bmatrix} B_k \\ D_k \end{bmatrix} r_k
$$

$$
= r'_k r_k \tag{4.12}
$$

Substituting from Eqs.(4.12) and (4.11c) into the objective function (4.10) and from Eqs.(4.11a, 4.11b) into the dynamics (4.1a) and the observation equation (4.1b), we get

$$
\min_{\hat{x}} \max_{r, x_0} \frac{1}{2} \left(\sum_{k=1}^{N} e'_k e_k - \gamma^2 \sum_{k=0}^{N-1} r'_k r_k - \gamma^2 (x_0 - \hat{x}_0)' P_0^{-1} (x_0 - \hat{x}_0) \right)
$$

$$\tag{4.13}$$

subject to

$$
\begin{aligned}
x_{k+1} &= A_k x_k + B_k r_k \\
e_k &= M_k (x_k - \hat{x}_k) \\
y_k &= C_k x_k + D_k r_k
\end{aligned} \tag{4.14}
$$

which is exactly the game theoretic formulation of Section 3.2.

4.3   Extension to Systems with Modeling Uncertainty

Figure 4.3 shows the familiar input/output representation of a nominal plant P with modeling uncertainties Δ and an estimator F. Here $\epsilon \equiv [\epsilon_0, \ldots, \epsilon_{N-1}]$ and $\eta \equiv [\eta_0, \ldots, \eta_{N-1}]$ represent the signals connecting the nominal plant and the perturbation, while $r, x_0, \hat{x}_0, y$, and e have the same definitions as in Section 4.2. Define $d_k = [r'_k \ \eta'_k]'$, where r_k is the same as in the previous section. Then, the state space representation is exactly the same as that of Eq.(3.29):

$$
\begin{aligned}
x_{k+1} &= A_k x_k + G_k r_k + Q_k \eta_k \\
&= A_k x_k + B_k d_k \\
\epsilon_k &= S_k x_k + T_{rk} r_k + T_{\eta k} \eta_k \\
&= S_k x_k + T_k d_k \\
e_k &= M_k (x_k - \hat{x}_k) \\
y_k &= C_k x_k + E_k r_k + R_k \eta_k \\
&= C_k x_k + D_k d_k
\end{aligned}
$$

$$\tag{4.15a}$$
$$\tag{4.15b}$$
$$\tag{4.15c}$$
$$\tag{4.15d}$$

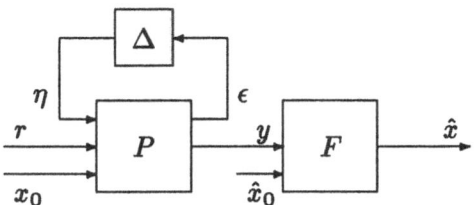

Figure 4.1: General representation of robust estimation problem

To extend the results of Section 4.2, define the optimization problem:

$$\min_{\hat{x}} \mathcal{L} \tag{4.16}$$

$$\text{subject to} \quad \text{Eqs.}(4.15a - 4.15d)$$

$$\|\eta\|^2 \leq \beta \|\epsilon\|^2 \quad \beta < 1 \tag{4.17}$$

The last constraint expresses the fact that the induced norm of Δ is bounded. As we did in Section 4.2, the objective function (4.16) can be rewritten to give the optimization problem

$$\min_{\hat{x}} \max_{(d,x_0)} \left(J - \theta^{-1} \Psi \right)$$

$$\text{subject to} \quad \text{Eqs.}(4.15a - 4.15d)$$

$$\|\eta\|^2 \leq \beta \|\epsilon\|^2 \quad \beta < 1$$

Associating the Lagrange multiplier θ_2^{-1} with the boundedness constraint on Δ, the optimization problem can be rewritten

$$\min_{\hat{x}} \max_{(r,\eta,x_0)} \sum_{k=1}^{N} e_k' e_k + \frac{\beta}{\theta_2} \sum_{k=0}^{N-1} \epsilon_k' \epsilon_k$$

$$- \left(\sum_{k=0}^{N-1} \frac{1}{\theta_2} \eta_k' \eta_k + \frac{1}{\theta} r_k' r_k \right)$$

$$- \left(\frac{1}{\theta} \|x_0 - \hat{x}_0\|_{P_0}^{-1} \right) \tag{4.18}$$

$$\text{subject to} \quad \text{Eqs.}(4.15a - 4.15d)$$

The above optimization problem is more general than that of Section 3.3. Specifically, if

$$\beta = \theta_2 = \theta$$

then, with $d_k = [\eta_k \; r_k]$, the objective function (4.18) becomes

$$\min_{\hat{x}} \max_{(d,x_0)} \left[\; \sum_{k=1}^{N} e_k' e_k + \sum_{k=0}^{N-1} \epsilon_k' \epsilon_k \right.$$

$$-\frac{1}{\theta} \left(\sum_{k=0}^{N-1} d_k' d_k \right.$$

$$\left. + \; \|x_0 - \hat{x}_0\|_{P_0}^{-1} \right) \right] \tag{4.19}$$

The above objective function becomes exactly the robust game theoretic problem of Section 3.3. This can be seen by comparing Eq.(4.19) and (3.32), and the associated constraints.

The above comparison shows that the robust risk sensitive estimation problem of Eq.(4.18) is equivalent to a game theoretic one more general than the problem of Chapter 3. This fact suggests another approach to dealing with disturbance and plant model uncertainties. If $\beta = \theta_2$, then we can vary the parameters θ and θ_2 independently in order to selectively fine tune the degree of robustness to both kinds of disturbances. Solving this problem is an avenue for future research.

4.4   Numerical Comparison of Error Density Functions

The error function J over a finite-time interval can be written as

$$J = e'e$$

where

$$e = [e_1', \ldots, e_N']'$$
$$e_k = M_k(x_k - \hat{x}_k)$$

If the density function of e is normal with mean zero and covariance Q, then it is shown in Appendix B that J can be expressed as the weighted sum of independent χ^2 random variables, where the weights σ_i are the singular values of $\mathcal{Q}$. Specifically, J has the same probability density function as

$$J \stackrel{d}{=} \sum_{k=1}^{N} \sigma_i t_k^2$$

where $\stackrel{d}{=}$ stands for equal in distribution, and the t_k^2's are each χ^2 random variable with one degree of freedom. The probability density function is derived in Appendix B.

With the stochastic interpretation, the Kalman filter, H_∞ optimal estimator, and more generally, the game theoretic estimator can be compared in terms of the probability density function of the quadratic estimation error. Steady-state frequency domain results such as those of Chapter 3 are also included in Figure 4.2 for comparison purposes.

Figure 4.3 shows the density function for the same plant and estimator designs used in Figure 4.2. Notice that with the risk sensitive or H_∞ optimal estimator, which is obtained by setting the parameter θ (γ) to its highest (lowest) possible value, the probability of a large value of J is lower, i.e., its density function has a thinner tail. This is because as θ increases, or equivalently γ decreases, more weight is placed on the higher moments, as can be clearly seen in the expansion of Eq.(4.6). The cost of reducing the tail with the H_∞ optimal estimator is a higher mean in J. Figures 4.2 and 4.3 show the tradeoff between using the H_∞ optimal estimator versus the Kalman filter. The H_∞ optimal estimator is robust in the sense that it minimizes the squared error for the worst-case bounded-energy noise. It also minimizes the weighted sum of higher moments of J given stochastic noise. As a result, the probability of large error is reduced. The expense is less rejection of high frequency noise, and a higher expected squared error. Therefore, it is best to use the most general form of the game theoretic estimator (or the equivalent risk sensitive estimator) and treat γ (or θ) as a design parameter that can be tuned for the objectives of a particular design.

Figure 4.4, which shows the same cost function as Figure 2.2, obviates the importance of considering modeling error in analyzing the performance of an estimator. In failure detection, for example, if a threshold is selected for a hypothesis test based on the residual of the Kalman filter, then the false alarm rate would be much higher than anticipated for the real, perturbed plant. The robust estimator is much less sensitive to this variation as can be seen in Figure 4.5.

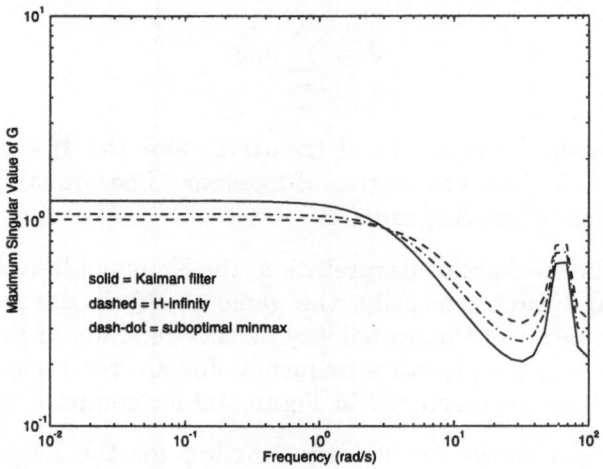

Figure 4.2: Frequency response from noise to estimation error.

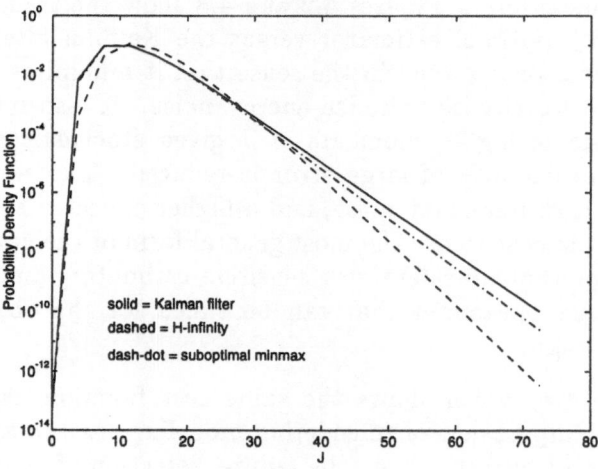

Figure 4.3: Probability density function of quadratic estimation error.

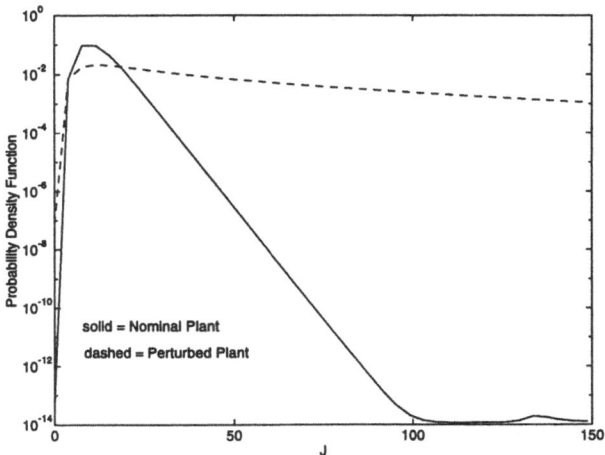

Figure 4.4: Probability density function comparison for Kalman filter.

Consider now the nominal and perturbed plants separately. Figure 4.6 shows the probability density function of the squared estimation error J with the nominal plant dynamics. Figure 4.7 shows the same function with the perturbed dynamics. These plots also show the large change in density function for both the Kalman filter and H_∞ optimal estimator with plant uncertainty. Figure 4.6 reveals that robustness may come at the expense of nominal performance.

4.5   Summary

In this chapter, we have shown that robust estimation has a stochastic interpretation in a general risk sensitive formulation. This interpretation gives additional insight into the meaning of robustness, as was seen by computing the probability density functions of the estimation errors.

First, in the absence of plant model uncertainty, the probability density function of the sum of squared estimation error had a thinner tail than that of the Kalman filter since, as a result of the exponential in the objective function of the risk sensitive formulation, higher order moments are included in the cost function. With the presence of plant

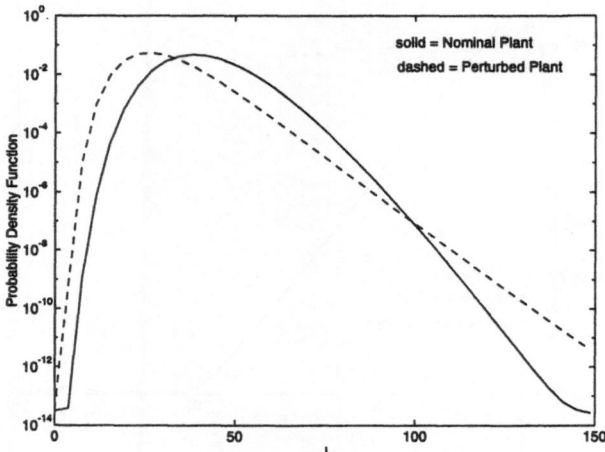

Figure 4.5: Probability density function comparison for robust estimator.

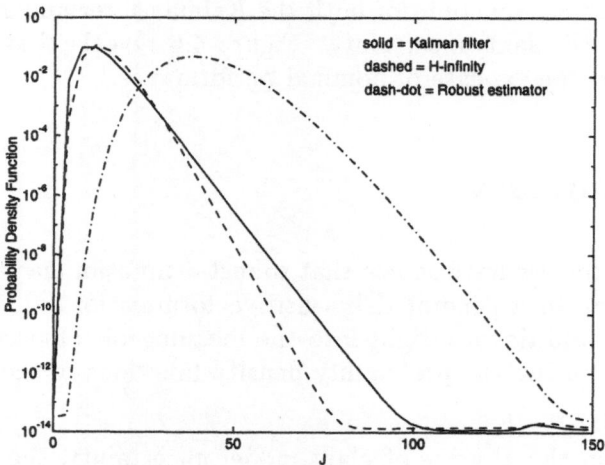

Figure 4.6: Probability density function with nominal plant.

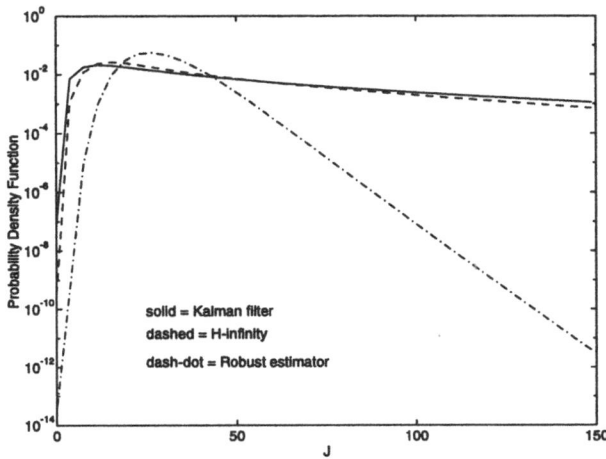

Figure 4.7: Probability density function with perturbed plant.

model uncertainties, robustness is illustrated by the fact that the output error density function of the nominal and perturbed systems are similar. Robustness, however, can come with a penalty of higher error variance.

We mention briefly another stochastic view of H_∞, suggested in [93], where the concept of stochastic gain is introduced. The stochastic gain is the supremum, with respect to a family of input disturbance probability density functions, of the steady-state expected value of the ratio of the energy norm of the system's output to the energy norm of the input disturbance. The goal is then to minimize the stochastic gain. Only systems with known plant models are considered in [93].

CHAPTER 5
ROBUST FAILURE DETECTION AND ISOLATION

5.1 Introduction

The basic objective of a fault detection and isolation methodology for dynamic systems is to detect failures and to identify, or isolate, the failed component. The most obvious method for automatic fault detection is the use of hardware redundancy, where measurements from multiple sensors are compared, and the existence of a failure is determined by implementing a voting mechanism. In many situations, however, hardware redundancy may not be possible or desirable, since it imposes a penalty in terms of volume, weight, etc. In other situations, such as with actuators, direct measurement is often not possible. In these cases, indirect measurements may be used to infer the component fault status using a model of the system. One method to analytically detect the existence of a failure is to look for anomalies in the plant's output relative to a model-based estimate of that output. Plant models, however, are generally incomplete and inaccurate. Moreover, these failure detection and isolation algorithms often assume a particular failure mode. These plant dynamics and failure mode modeling errors can either cause a high false alarm rate, or make it difficult to detect failures. Any robust detection and isolation test that is designed to overcome the problems associated with these modeling errors must be able to distinguish between model uncertainties and failures in order to avoid excessive false alarms or missed detections.

In this chapter, a methodology for the detection and isolation of failures in the presence of model uncertainties is developed. The objective is to design an algorithm that is insensitive to failure mode, noise and plant model uncertainties, but sensitive to the occurrence of a failure. As mentioned in Chapter 2, we will be concerned with failures in the plant's control and measurement systems, comprising actuators and sensors.

For robustness to failure mode uncertainty, the hypothesis underlying the algorithm assumes that a component failure is a sample path from a Gauss-Markov model, or equivalently, that the failure is the output of a shaping filter. Such a model embraces a large class of failures. This shaping filter is appended to the plant's dynamic model, giving an augmented linear system with plant and failure states. For robustness to noise and plant model uncertainties, the algorithm relies on the robust game theoretic or risk sensitive filter, derived in the earlier chapters, just as the likelihood ratio test relies on the Kalman filter to obtain a maximum likelihood estimate of the failure.

The statement of the robust detection and isolation problem is given in the next section. In Section 5.3, we derive a likelihood ratio test based on a Gauss-Markov failure model, assuming an accurate plant model. This derivation leads to an algorithm that is robust to failure mode and noise model uncertainties. In Section 5.4, we examine the effect of plant model uncertainties on this algorithm. In Section 5.5, we describe an algorithm that is robust to failure mode, noise and plant model uncertainties. A summary of the chapter appears in Section 5.6.

5.2   Problem Description

A general discussion of the problem is given in Section 5.2.1, where notation is also introduced. A formulation of the robust failure detection test is given in Section 5.2.2. For this problem to be well posed, a failure model must be assumed. This model is described in Section 5.2.3.

5.2.1   General Discussion and Notation

A failure detection test is a hypothesis test between a set of unfailed plants and a set of failed plants. Figure 5.1 illustrates the two hypotheses. The figure shows a general input/output representation of

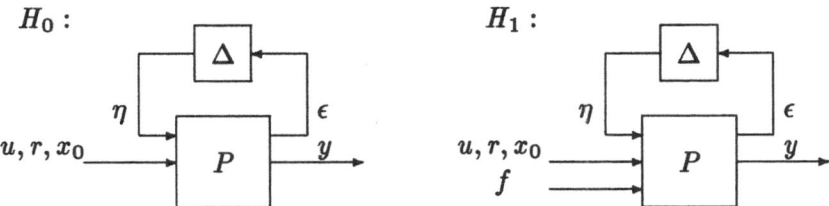

Figure 5.1: Hypothesis test for additive failures in the presence of model uncertainty

a nominal plant P with modeling uncertainty Δ. The vector u represents the known input to the plant, r represents the combined process and measurement noise, x_0 represents the initial state vector of the nominal plant, and y is the sensor measurement. The signals ϵ and η represent the interaction between the nominal plant and the perturbation Δ. Finally, f represents the failure signal. Our concern is for failures in the actuators and sensors of the plant. We will shortly see that such failures have the additive representation shown in Figure 5.1. If

$$\mathcal{P} = \{(P, \Delta) \mid \Delta \in \boldsymbol{\Delta}\} \tag{5.1}$$

represents the nominal plant P and the set $\boldsymbol{\Delta}$ of all permissible perturbations due to modeling errors, and $\mathcal{P}_f$ represents the same set of plants with a failure injected, then the two hypotheses between which a detection test must choose are:

$$H_0 : \quad (P, \Delta) \in \mathcal{P}$$
$$H_1 : \quad (P, \Delta) \in \mathcal{P}_f$$

The objective of a failure detection test is to determine whether a fault exists somewhere in the plant. By contrast, a failure isolation test aims to identify the failed components. One approach towards failure isolation is to first perform a detection test. Then, if a failure is detected, a more elaborate isolation hypothesis test can be used. For any pair of failed plants (i, j), the isolation test can be described as a choice between as

$$H_i : \quad (P, \Delta) \in \mathcal{P}_i$$
$$H_j : \quad (P, \Delta) \in \mathcal{P}_j$$

where $\mathcal{P}_i$ ($\mathcal{P}_j$) denote the set of all possible plants with a failure injected into the ith (jth) component.

Consider now an interval of interest, say $[k_0, ..., k_0 + N]$. The following notation is used to denote the various signals:

$$\begin{aligned}
r &\equiv [r_{k_0}, ..., r_{k_0+N}] \\
\eta &\equiv [\eta_{k_0}, ..., \eta_{k_0+N}] \\
\epsilon &\equiv [\epsilon_{k_0}, ..., \epsilon_{k_0+N}] \\
y &\equiv [y_{k_0}, ..., y_{k_0+N}] \\
f &\equiv [f_{k_0}, ..., f_{k_0+N}]
\end{aligned}$$

Each of the above signals is a matrix, whose kth column represents the value of all the signal components at time step k, and whose rows are the time history of various components of the signal. Thus, f_k, $k = k_0, ..., k_0 + N$, denotes the vector of all the failures at time k:

$$f_k = \begin{bmatrix} f_{1k} \\ \cdots \\ f_{\mathcal{M}k} \end{bmatrix}$$

By contrast, we use the notation $f_{i,\cdot}$, $i = 1, ..., \mathcal{M}$ to denote the ith row of f, or the time history of the ith failure element:

$$f_{i,\cdot} = \begin{bmatrix} f_{ik_0}, & \cdots & f_{i,k_0+N} \end{bmatrix} \tag{5.2}$$

The ℓ_2 norm of the input disturbance r is given by

$$\|r\| \equiv \left(\sum_{k=k_0}^{k_0+N} r_k' r_k \right)^{\frac{1}{2}} \tag{5.3}$$

The ℓ_2 norms of η, ϵ, and y are similarly defined. The norm of the failure signal f can be slightly more general, since it can include weights, as will be seen later in the chapter. In addition, $\|x_0 - \hat{x}_0\|_{P_0^{-1}}$ represents the weighted Euclidean norm of the initial estimation error.

The disturbances can be seen as either deterministic or stochastic. For a deterministic model, r has a bounded ℓ_2 norm, while $x_0 - \hat{x}_0$ is assumed to have a bounded weighted Euclidean norm. For a stochastic model, r is a white Gaussian noise sequence with unit variance and $x_0 - \hat{x}_0$ is Gaussian with mean zero and covariance P_0. Furthermore, the r_k's are uncorrelated with the the initial error $x_0 - \hat{x}_0$. Because of the risk sensitive interpretation of game theoretic estimation described in Chapter 4, the robust FDI algorithm developed is applicable to either a stochastic or deterministic setting. These two interpretations help give insight into the design and analysis of the algorithm. Furthermore, Δ is the set of perturbations whose induced 2-norm is given by

$$\Delta \equiv \left\{ \Delta \mid \|\Delta\|_{i2} \equiv \sup_{\epsilon \neq 0} \frac{\|\eta\|}{\|\epsilon\|} < 1 \right\} \qquad (5.4)$$

The developments in this chapter are for linear plants, time-varying and time-invariant. If the stochastic setting is assumed, then the set Δ will be restricted to linear perturbations only. For the deterministic view, no such assumption on the perturbation is needed.

5.2.2   Problem Formulation

Let x_k be the state of the dynamic plant P at time k, and y_k be the observations. Note that x_k can consist of the nominal plant's state, as well as any additional state or states introduced as part of the uncertainty model, if needed. The no-failure (H_0) and failure (H_1) hypotheses for the detection test over an interval $[k_0, k_0 + N]$ can now be formally introduced:

$$H_0 \ : \ \begin{aligned} x_{k+1} &= Ax_k + Q\eta_k + Br_k + Uu_k \\ \epsilon_k &= Sx_k + T\eta_k \\ y_k &= Cx_k + R\eta_k + Dr_k + Wu_k \end{aligned} \qquad (5.5)$$

$$H_1 \ : \ \begin{aligned} x_{k+1} &= Ax_k + Q\eta_k + Br_k + Uu_k + Ff_k \\ \epsilon_k &= Sx_k + T\eta_k \\ y_k &= Cx_k + R\eta_k + Dr_k + Wu_k + Lf_k \end{aligned} \qquad (5.6)$$

with initial condition $x = \hat{x}_0$, as given before. In Eqs.(5.5-5.6), the perturbation's output signal η enters the plant through the matrices Q and R, while S and T represent the plant's input into the uncertainty. As stated in Chapter 3, this formulation can represent a large class of uncertainties. These include parametric uncertainties, such as perturbations in the matrices A, B, C, D, U and W, as well as nonparametric uncertainties, such as unmodeled dynamics. In Section 3.3.1, we showed how parametric uncertainties can be represented. In Appendix D, a frequency domain approach to modeling uncertainties is presented.

Each f_{ik} represents the failure of a control or measurement channel at time k. The matrices F and L describe the way the control command and measurement failures enter into the system. In the equations for the failure hypothesis, the failures enter the system additively. Though an additive representation is not always applicable to failures in the plant's structure, such as an abrupt change in an element of the plant's A matrix, it is realistic for a wide range of actuator and sensor failures.

Note that the time at which the failure occurs does not figure in the hypothesis test. That is, the failure is assumed either to exist or not to exist for the entire interval of observations. For the sake of notational compactness, the problem statement, and the subsequent development, is presented for a time-invariant system, although both are applicable to time-varying systems as well, with A_k replacing A, etc.

5.2.3   The Failure Model

For the hypothesis test of Eqs.(5.5-5.6), we need to specify our assumptions about the failure f. One possibility is to assume a step failure of unknown magnitude, that is, $\forall i = 1, ..., \mathcal{M}$,

$$f_{ik} = \nu_i \ \forall k \in [k_0, ..., k_0 + N] \tag{5.7}$$

where ν_i is a constant. Such a model does not allow for other possible failure modes, such as ramp failures, sinusoidials, etc. To achieve robustness to failure mode uncertainty, one possible model, already mentioned in Chapter 2, is

$$f_{ik} = \sum_{j=1}^{J_i} \phi_{jk} \varsigma_{ij} \qquad \forall i \in [1, \mathcal{M}]$$

$$\forall k \in [k_0, k_0 + N] \tag{5.8}$$

where the ϕ_{jk}'s form the basis of a function subspace, and ς_{ij} are the unknown but constant projections. This failure representation is suggested in [42], page 338 . One possible expansion is the Haar basis [102], which is appropriate for abrupt changes. Other examples are orthogonal polynomials such as the Laguerre or Legendre basis, used in [47].

For our hypothesis test, we will assume that the failure model is a shaping filter of the form

$$\varphi_{k+1} = A_f \varphi_k + B_f \vartheta_k \tag{5.9}$$

$$f_k = C_f \varphi_k \tag{5.10}$$

where A_f is a *stable* filter. The above model can be time-varying or time-invariant. It can be viewed as deterministic or stochastic. For a deterministic model, ϑ has a bounded ℓ_2 norm, and φ_0 is assumed to have a bounded Euclidean norm. For a stochastic model, Eq.(5.9) is a Gauss-Markov model, where ϑ is a white noise sequence with unit variance and φ_0 is Gaussian with mean zero and covariance P_{φ_0}.

The step failure model of Eq.(5.7) can be seen as a special case of the shaping filter (5.9-5.10). In this case, $A_f = I$, $B_f = 0$, $C_f = I$, and $P_{\varphi_0} = \infty$. The choice of an infinite initial covariance is dictated by the fact that no prior information on the failure is assumed. If the Laguerre or Legendre basis is used in Eq.(5.8), then their coefficients can be generated by a stable shaping filter of the form of Eqs.(5.9-5.10) as well [113].

For our choice of failure model, we will assume that each row $f_{i,\cdot}$ (see Eq.5.2) of the failure signal f, which represents the ith modeled failure, is given by a first order shaping filter:

$$\begin{aligned} \varphi_{i,k+1} &= a_{f_i}\varphi_{ik} + b_{f_i}\vartheta_{ik} \\ f_{ik} &= \varphi_{ik} \end{aligned} \tag{5.11}$$

where f_{i0} has a mean of 0 and some variance or, alternatively, f_{i0} has a bounded ℓ_2 norm. The above model is also a special case of Eqs.(5.9-5.10). Specifically,

$$A_f = \begin{bmatrix} a_{f_1} & & \\ & \ddots & \\ & & a_{f_\mathcal{M}} \end{bmatrix} \tag{5.12}$$

$$B_f = \begin{bmatrix} b_{f_1} \\ \vdots \\ b_{f_\mathcal{M}} \end{bmatrix} \tag{5.13}$$

$$C_f = I \tag{5.14}$$

$$\hat{f}_0 = 0 \tag{5.15}$$

The one-dimensional model of Eq.(5.11) is desirable for several reasons. First, each failure is represented by one state only, which is computationally simpler. Second, there are only two parameters to determine, the bandwidth a_{f_i} and the amplitude b_{f_i}/a_{f_i}, for which most users have a good intuitive feel. The higher the bandwidth, for instance, the larger the class of failures under consideration. A failure signal whose frequency content is higher than the bandwidth would be treated as noise. We will return to the choice of bandwidth and amplitude in Sections 5.3 and 5.5.2. Another advantage is that the model can be used with a steady-state filter, whenever such a filter is desirable.

Note that any model of the form of Eqs.(5.9 - 5.10) is an idealization of the failure process, used for the sake of convenience. Most failures are bounded and occur abruptly. For instance, a rudder deflection cannot exceed a certain maximum angle, but the output of a Gauss-Markov model has no deterministic bound. Moreover, a low-pass filter, or most expansions, are not exact models for an abrupt change. The representation of Eqs.(5.9-5.10), however, is a tractable linear state-space model that can be used to approximate a wide class of failures.

Finally, the state dynamic equation of the failed hypothesis (5.6) can be augmented with the failure model of Eqs.(5.9-5.10) to give

$$\begin{bmatrix} x_{k+1} \\ \varphi_{k+1} \end{bmatrix} = \begin{bmatrix} A & FC_f \\ 0 & A_f \end{bmatrix} \begin{bmatrix} x_k \\ \varphi_k \end{bmatrix} + \begin{bmatrix} Q \\ 0 \end{bmatrix} \eta_k + \begin{bmatrix} B & 0 \\ 0 & B_f \end{bmatrix} \begin{bmatrix} r_k \\ \vartheta_k \end{bmatrix}$$
$$+ \begin{bmatrix} U \\ 0 \end{bmatrix} u_k \qquad (5.16)$$

with initial estimate

$$\begin{bmatrix} x_0 \\ f_0 \end{bmatrix} = \begin{bmatrix} \hat{x}_0 \\ 0 \end{bmatrix} \qquad (5.17)$$

If a stochastic setting is assumed, then the initial error has a mean of zero and a covariance given by

$$P_0 = \begin{bmatrix} \check{P}_0 & 0 \\ 0 & P_{\varphi_0} \end{bmatrix} \qquad (5.18)$$

while with a deterministic setting, the initial estimation error has a weighted Euclidean norm, with weight given by P_0^{-1}. A joint bound is assumed on the norm of the disturbances and initial error

$$\|\phi_0\| + \|r\|^2 + \|x_0 - \hat{x}_0\|_{P_0^{-1}}^2 + \|\vartheta\|^2 < 1 \qquad (5.19)$$

Again, the bound of one can be changed by rescaling the matrices in Eq.(5.16). Finally, the associated observation equation is

$$y_k = \begin{bmatrix} C & LC_f \end{bmatrix} \begin{bmatrix} x_k \\ \varphi_k \end{bmatrix} + \begin{bmatrix} D & R & W & 0 \end{bmatrix} \begin{bmatrix} r_k \\ \eta_k \\ u_k \\ \vartheta_k \end{bmatrix} \qquad (5.20)$$

The ability of any FDI algorithm that relies on the failure model (5.11) to detect failures is tied to the observability of the augmented system of Eq.(5.16). That is, if

$$\left(\begin{bmatrix} A & FC_f \\ 0 & A_f \end{bmatrix}, \begin{bmatrix} C & LC_f \end{bmatrix} \right)$$

is observable, then, at least in steady-state, the algorithm's filter will be able to provide an estimate of the failure signal f. The quality of the failure estimate determines our ability to isolate the failed component, or components. In the absence of plant model uncertainty, the observability condition just mentioned is therefore important. The speed of detection, or isolation, however, is determined by the bandwidth of the mapping between the failure and its estimate.

In the presence of model uncertainty, the problem is more difficult, as we need to be insensitive to the uncertainty, while remainig sensitive to the failure estimate, or at least sensitive enough to determine which component failed. For this, the robust FDI algorithm relies on a robust filter to generate a residual that is sensitive to failure, but insensitive to uncertainty. The robust filter is designed for the augmented system of Eqs.(5.16-5.20). The residual used by the decision test is a function of the failure estimate. This FDI algorithm can be seen as an extension of likelihood ratio tests, as we will show in the next three sections.

5.3   A Likelihood Ratio Test with Robustness Properties

The objective in this section is to derive the likelihood ratio for the detection and isolation tests of the previous section, assuming the Gauss-Markov failure model (5.11), and accurate knowledge of the plant and model statistics. As a result, $Q = R = S = T = 0$ in the two hypotheses (5.5) and (5.6). By showing that this test relies on a risk sensitive or game theoretic estimator, we demonstrate robustness to failure mode and noise model uncertainties only. The algorithm derived in Section 5.5 is an extension of this test to the case where plant uncertainties are present, and reduces to it in the absence of such uncertainties.

The goal is to show that the decision statistic is a quadratic function of the least squares estimate of the failure, given the observations, as is the case with any likelihood ratio test in the Gaussian context. For a nominal model, the *weighted* likelihood ratio over an interval $[k_0, ..., k_0 + N]$ for the failure detection test is

$$
\begin{aligned}
\Lambda^s &= \frac{E_{\underline{f}}[p(Y \mid H_1)]}{p(Y \mid H_0)} \\
&= \frac{\int_{-\infty}^{+\infty} p(Y \mid H_1, \underline{f}) p(\underline{f} \mid H_1) d\underline{f}}{p(Y \mid H_0)}
\end{aligned}
\tag{5.21}
$$

The superscript s will be explained shortly. The vector Y represents the observations over the entire interval $[k_0, k_0 + N]$ stacked into one column:

$$
\begin{aligned}
Y &= \begin{bmatrix} y_{k_0} \\ \cdots \\ y_k \\ \cdots \\ y_{k_0+N} \end{bmatrix} \\
&= [y'_{k_0}, ..., y'_{k_0+N}]'
\end{aligned}
$$

and, similarly, $\underline{f}$ is simply the signal f with its columns f_k, $k = k_0, ..., k_0 + N$ stacked into one column:

$$
\underline{f} = [f'_{k_0}, ..., f'_{k_0+N}]'
$$

The vector $\underline{f}$ has a mean of zero and a covariance Σ_f that can be derived from Eq.(5.11), seen as a Gauss-Markov model (see Appendix B). For the entire interval, this test is given as

$$
H_0 \; : \; Y \; = \; Y_0 \tag{5.22}
$$

$$
H_1 \; : \; Y \; = \; Y_0 + \mathcal{G}\underline{f} \tag{5.23}
$$

where Y_0 is the vector of observations in the absence of failure, and the matrix $\mathcal{G}$ represents the projection of the failure process onto the observations

$$
\mathcal{G} = \begin{bmatrix}
L & & & & \\
CF & L & & & \\
\cdots & \cdots & \cdots & & \\
CA^{k-k_0-1}F & CA^{k-k_0-2}F & \cdots & L & \\
\cdots & \cdots & \cdots & & \\
CA^{N-1}F & CA^{N-2}F & \cdots & \cdots & CF & L
\end{bmatrix}
\tag{5.24}
$$

The test therefore reduces to the classical problem of detecting a stochastic signal, f or $\underline{f}$, in colored noise, Y_0. The density function of the observation under each of the two hypotheses is given by

$$H_0 \quad : \quad Y \sim \mathcal{N}(0, \Sigma_0) \tag{5.25}$$

$$H_1 \quad : \quad Y \sim \mathcal{N}\left(0, \Sigma_0 + \mathcal{G}\Sigma_{\underline{f}}\mathcal{G}'\right) \tag{5.26}$$

where Σ_0 is the covariance of the observation under the null hypothesis, computed from the model of Eqs.(5.16 -5.20), together with the observation equation in (5.5). Note that $\underline{f}$ and Y_0 are independent, which allows us to add the covariances in Eq.(5.26). In terms of Y, the likelihood ratio of Eq.(5.21) is given by

$$\Lambda^s = \text{Constant} \times \frac{e^{-Y'\left(\Sigma_0 + \mathcal{G}\Sigma_{\underline{f}}\mathcal{G}'\right)^{-1}Y}}{e^{-Y'\Sigma_0^{-1}Y}} \tag{5.27}$$

Taking the log of the above expression, we get, apart from an additive constant,

$$
\begin{aligned}
\mathcal{D}^s \quad &= \quad \log \Lambda^s \\
&= \quad -Y'\left[\left(\mathcal{G}\Sigma_{\underline{f}}\mathcal{G}' + \Sigma_0\right)^{-1} - \Sigma_0^{-1}\right]Y
\end{aligned}
\tag{5.28}
$$

The above ratio can be expressed in terms of the maximum a posteriori (MAP) estimate of $\underline{f}$, given the observation sequence Y. In the linear Gaussian context, the MAP estimate is also the *smoothed* minimum variance estimate, to be denoted by $\hat{\underline{f}}^s$. The superscript s on the ratios Λ^s and $\mathcal{D}^s$ have been added to emphasize the fact that these depend on a smoothed estimate of the failure. To show this, first express this estimate as a function of the observations:

$$
\begin{aligned}
\hat{\underline{f}}^s \quad &= \quad E\left(\underline{f} \mid Y\right) \\
&= \quad \left(\mathcal{G}'\Sigma_0^{-1}\mathcal{G} + \Sigma_{\underline{f}}^{-1}\right)^{-1}\mathcal{G}'\Sigma_0^{-1}Y \\
&= \quad \Sigma_{\underline{f}|Y}\mathcal{G}'\Sigma_0^{-1}Y
\end{aligned}
\tag{5.29}
$$

where

$$\Sigma_{\underline{f}|Y} = \left(\mathcal{G}'\Sigma_0^{-1}\mathcal{G} + \Sigma_{\underline{f}}^{-1}\right)^{-1} \tag{5.30}$$

is the a posteriori covariance of the failure given the observation. To obtain Eq.(5.29), first derive the joint density of f and $Y = Y_0 + \mathcal{G}f$, then apply Theorem A.1. Using the matrix inversion lemma, we have

$$
\begin{aligned}
\left(\mathcal{G}\Sigma_{\underline{f}}\mathcal{G}' + \Sigma_0\right)^{-1} \quad &= \quad \Sigma_0^{-1} - \Sigma_0^{-1}\mathcal{G}\left(\mathcal{G}'\Sigma_0^{-1}\mathcal{G} + \Sigma_{\underline{f}}^{-1}\right)^{-1}\mathcal{G}'\Sigma_0^{-1} \\
&= \quad \Sigma_0^{-1} - \Sigma_0^{-1}\mathcal{G}\Sigma_{\underline{f}|Y}\mathcal{G}'\Sigma_0^{-1}
\end{aligned}
\tag{5.31}
$$

Substituting for the left-hand side of the above equation into Eq.(5.28), we get

$$\mathcal{D}^s = Y'\Sigma_0^{-1}\mathcal{G}\Sigma_{\underline{f}|Y}\mathcal{G}'\Sigma_0^{-1}Y \qquad (5.32)$$

Comparing the above equation with Eq.(5.29), we see that the log likelihood ratio test can be expressed as

$$\mathcal{D}^s = \underline{\hat{f}}^{s'}\Sigma_{\underline{f}|Y}^{-1}\underline{\hat{f}}^s \underset{<}{\overset{>}{\gtrless}} \text{threshold} \qquad (5.33)$$

The above expression shows explicitly that likelihood ratio detection tests over an interval make use of a smoothed rather than a filtered estimate. This fact is not made obvious in [114] because the failure is assumed to be a jump whose magnitude does not vary with time. As such, the filtered estimate of the failure at the end of the interval $[k_0, k_0 + N]$, which can be obtained using a Kalman filter, is also a smoothed estimate.

Notice that $\underline{\hat{f}}^s$ is a fixed-interval smoothed estimate, and it can be obtained using a backward and forward, or causal, filter, as explained in Section A.8.1. Recall also that in Section 3.5, we have shown that minimum variance fixed-interval smoothing is equivalent to game theoretic fixed-interval smoothing. For this reason, one can consider the estimate $\underline{\hat{f}}^s$ robust to noise model uncertainty!

For convenience, it is often desirable to use a causal version of Eq.(5.33). To do so, we replace the smoothed estimate of each element of the failure vector by the filtered estimate. That is, instead of using a smoother based on Eqs.(5.16-5.20), we use a game theoretic or minmax filter based on the same equation. Recall from Chapters 3 and 4 that the minmax filter is parametrized by a parameter γ (or θ). If we set γ to infinity (or θ to zero), we have the Kalman filter. On the other hand, if we set γ (θ) to its minimum (maximum) possible value, we have the H_∞ filter. Intermediate values of these parameters trade off average and worst case noise performance. Define

$$\hat{f}^s = \left[\hat{f}_{k_0}^s, ..., \hat{f}_k^s, ..., \hat{f}_{k_0+N}^s \right]$$

That is, $\hat{f}^s$ is simply $\underline{\hat{f}}^s$ rearranged in the same way as f itself. Then,

$$\hat{f}_k^s = E\left(f \mid y_{k_0}, ..., y_{k_0+N}\right) \quad \forall k \in [k_0, ..., k_0 + N]$$

is replaced by $\hat{f}_k^c$, the solution to the following risk sensitive estimation problem

$$\min_{\hat{f}} \theta^{-1} \log E\left(e^{\theta J}\right) \qquad (5.34)$$

where

$$J \equiv \frac{1}{2} \sum_{k=1}^{k_0+N} e_k' e_k \tag{5.35}$$

$$\begin{aligned} e_k &\equiv M_k(f_k - \hat{f}_k) \\ &= M_k C_f (\varphi_k - \hat{\varphi}_k) \end{aligned} \tag{5.36}$$

subject to the dynamic constraints of Eqs.(5.16)-(5.18), rewritten below, with the plant uncertainty matrix Q and R set to zero

$$\begin{bmatrix} x_{k+1} \\ \varphi_{k+1} \end{bmatrix} = \begin{bmatrix} A & FC_f \\ 0 & A_f \end{bmatrix} \begin{bmatrix} x_k \\ \varphi_k \end{bmatrix} + \begin{bmatrix} B & 0 \\ 0 & B_f \end{bmatrix} \begin{bmatrix} r_k \\ \vartheta_k \end{bmatrix} + \begin{bmatrix} U \\ 0 \end{bmatrix} u_k \tag{5.37}$$

The initial estimate and error covariances are

$$\begin{bmatrix} x_0 \\ f_0 \end{bmatrix} = \begin{bmatrix} \hat{x}_0 \\ 0 \end{bmatrix} \tag{5.38}$$

and

$$P_0 = \begin{bmatrix} \check{P}_0 & 0 \\ 0 & P_{\varphi 0} \end{bmatrix} \tag{5.39}$$

The associated observation equation is

$$y_k = \begin{bmatrix} C & LC_f \end{bmatrix} \begin{bmatrix} x_k \\ \varphi_k \end{bmatrix} + \begin{bmatrix} D & W & 0 \end{bmatrix} \begin{bmatrix} r_k \\ u_k \\ \vartheta_k \end{bmatrix} \tag{5.40}$$

Now, if we drop the time-correlation matrix blocks from the covariance matrix $\Sigma_{f|Y}$, it reduces to a block diagonal matrix. As a result we have a recursive decision function for the interval window $[k_0, k + 0 + N]$,

$$\mathcal{D}_{k_0+N}^c = \sum_{k=k_0}^{k=k_0+N} \|\hat{f}_k^c\|_{\Sigma_k^{-1}}^2 \underset{<}{\overset{>}{\gtrless}} \text{threshold} \tag{5.41}$$

If the model of Eq.(5.11) for the failure is used, then the parameters a_f and b_f can be chosen so as to obtain rapid and accurate tracking of the failure. This can be done by choosing the steady-state gain of the transfer function $T_{\hat{f}_c f}$ between the input failure and the failure estimate to be close to unity for a large bandwidth. That is,

$$| T_{\hat{f}_c f}(\omega) | \simeq 1 \quad \forall \omega < \omega^* \qquad (5.42)$$

The goal is to have a filter that is as fast as possible, without causing a large false alarm probability by having the disturbances pass as a failure. The frequency ω^*, as well as the elements of A_f and B_f, are therefore determined by trial and error with Eq.(5.42) in mind.

Note that we have derived a family of failure detection and isolation tests applicable to a large family of failures, parametrized by the parameter θ! Thanks to this parameter, these tests are robust to noise model uncertainty, but they assume perfect knowledge of the plant model. In the next section, we analyze the consequences on the test's performance when this assumption is not valid.

5.4   Likelihood Ratio Tests and Plant Uncertainties

Given Gaussian noise, and no plant dynamic modeling uncertainty, the failure estimate is Gaussian with zero mean and covariance Σ_0. The decision function

$$\mathcal{D} = \underline{\hat{f}}' \Sigma_0^{-1} \underline{\hat{f}}$$

therefore has a χ^2 distribution with degrees of freedom equaling the length of the smoothed or filtered estimate $\underline{\hat{f}}$. Plant dynamic uncertainty will degrade the Kalman filter (smoother) estimate producing a covariance Σ different, and usually larger than the nominal value Σ_0. In addition, with a known plant model, known deterministic inputs (such as actuator deflections) are propagated through the plant model and nominally have no effect on the estimation error. With modeling uncertainty, these deterministic inputs can produce large biases in the failure estimate, significantly degrading performance.

The effects of modeling uncertainty are illustrated in Figure 5.2 for a two element Gaussian vector. The covariance can be represented by an ellipse with major and minor axes formed by the eigenvectors of the covariance matrix with lengths given by the eigenvalues. Modeling uncertainty will tend to increase and distort the covariance and non-zero mean inputs will produce non-zero mean outputs. Figure 5.3 shows

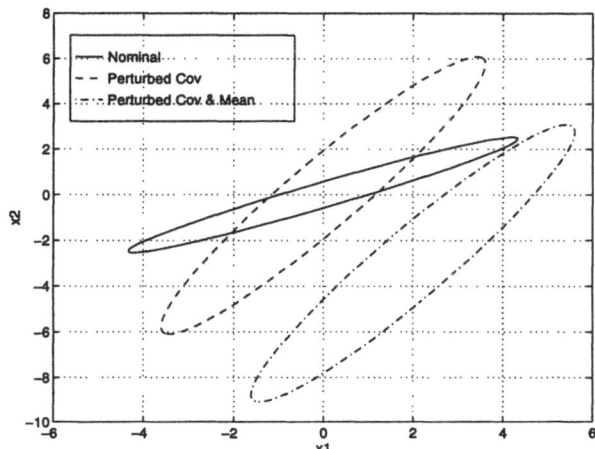

Figure 5.2: Effect of model uncertainty on the covariance of two element Gaussian vector.

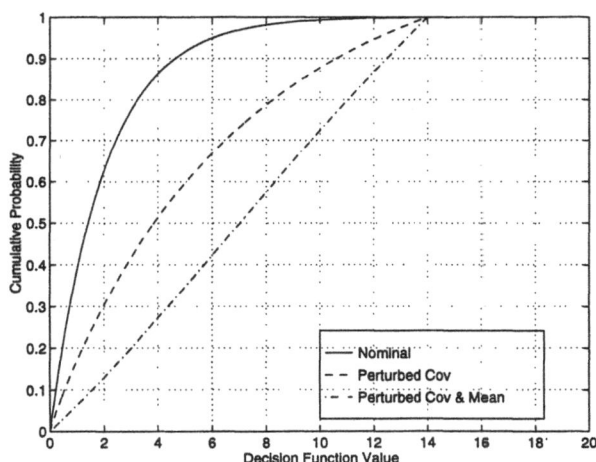

Figure 5.3: Effect of model uncertainty on decision function.

the effect of these distortions on the decision function for a hypothetical example.

In general, the decision function can be represented as a weighted sum of χ^2 variables. Given $\underline{f}$ with a Gaussian density with mean m_f and covariance Σ, the decision function can be written as

$$\mathcal{D} = Z' \Sigma^{1/2} \Sigma_0^{-1} \Sigma^{1/2} Z$$

where

$$Z = \Sigma^{-1/2} \hat{\underline{f}}$$

The decision function can be expressed as

$$\mathcal{D} \stackrel{d}{=} \sum_{j=1}^{\mathcal{M}} \lambda_j Z_j^2$$

where $\stackrel{d}{=}$ stands for equal in distribution, λ_j are the eigenvalues of $\Sigma_0^{-1/2} \Sigma^{1/2}$, and the Z_j's are independent Gaussian random variables with unity variance. The decision function is therefore the weighted sum of non-central χ^2 random variables with one degree of freedom.

Modeling uncertainty in the plant dynamics can produce a large degradation in performance. Simply ignoring the uncertiainty can lead to excessive false alarms. Accounting for the uncertainty by just increasing thresholds can lead to missed detections. Ideally, the mean and covariance of the failure estimates should be made as small as possible (with no failures) over the range of modeling uncertainty. In addition, the decision function $\mathcal{D}$ would ideally be formed using the actual covariance Σ. Unfortunately, there are as many possible covariances as there are possible plant models given the range of modeling perturbations. The selection of the normalization matrix that produces the best performance remains an open research question.

5.4.1   Examples: Underwater Vehicle with Model Uncertainty

To illustrate some of the points made in the above discussion, a likelihood ratio test was designed for the detection of failures in the stern plane (pitch control channel) of an underwater vehicle operating in open loop mode. The vehicle pitch dynamics have three states: depth rate, pitch angle, and pitch rate. There is one control input, the stern

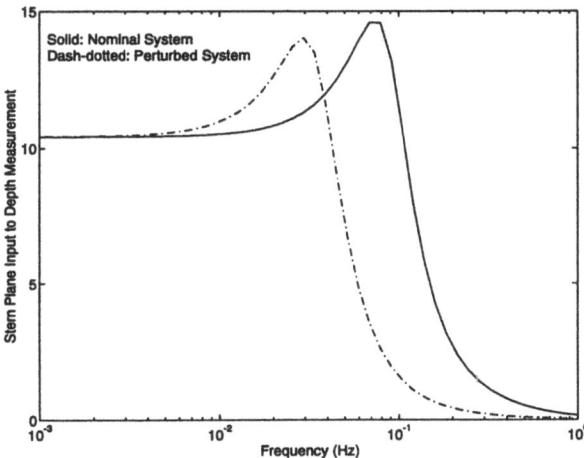

Figure 5.4: Magnitude of frequency response from the stern plane command to depth for two different configurations of the underwater vehicle.

plane command, and one disturbance input, a pitch rate torque. The measured variables are pitch and pitch rate. The vehicle has four fins, two horizontal, and two vertical. Only the horizontal fins, which execute the stern plane commands, affect the pitch dynamics. When one of these two horizontal fins fails, this command is affected.

The same likelihood ratio test design is to be used with two different configurations of the ARPA unmanned underwater vehicle (UUV). The two vehicle configurations differ in their mass, length and inertia. The linearized model dynamics for each configuration are given in Appendix D. Figures 5.4 and 5.5 illustrate how different the two models are. The plots represent, respectively, the magnitude Bode plot for the transfer function from the stern plane command to two different measurements, the depth and the pitch angle, for each vehicle configuration.

The Kalman filter design is based on the linearized model of the nominal vehicle configuration. In addition to the plant states, there is one state that represents failures in the stern plane command. The detection function of the likelihood ratio test is simply the square of the failure state estimate given by the Kalman filter. Isolation between the two fins for this example is simple, using the roll measurement.

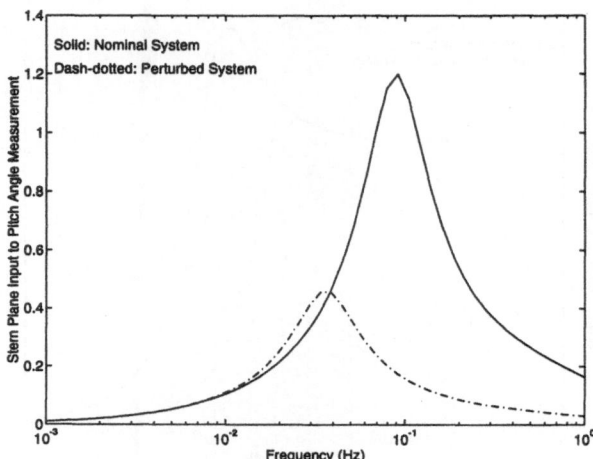

Figure 5.5: Magnitude of frequency response from the stern plane command to pitch angle for two different configurations of the underwater vehicle.

For the two examples given below, a steady-state Kalman filter is used.

Straight and Level Cruise

For this example, it is assumed that the vehicle is at steady-state in straight and level cruise, at a constant velocity, and that there is no significant control input, so that $\sum_{k=k_0}^{k_0+N} u_k \approx 0$ (the control simply cancels out the effects of process noise). Figure 5.6 compares the likelihood ratio test's probability of detection vs. probability of false alarm for a single implementation of the test. The results in the figure are for a ramp failure with a slope of .33 deg/sec and a maximum amplitude of 5 degrees. For a fixed probability of false alarm, say 10^{-6}, the probability of detection for the nominal system, as shown by the solid curve, is greater than .95. For the perturbed system, however, the corresponding probability of detection is less than 10^{-5}, as the dotted curve shows.

Figure 5.7 shows the time response of the nominal likelihood ratio detection function in the presence of a ramp failure injected at time $t = 50$ seconds. For this failure, a fin is deflected from 0 to 5 degrees

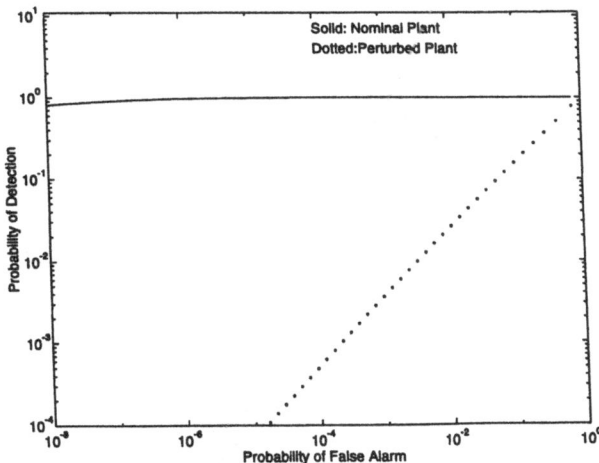

Figure 5.6: Probability of detection vs. Probability of false alarm with a ramp failure for the nominal and perturbed plants

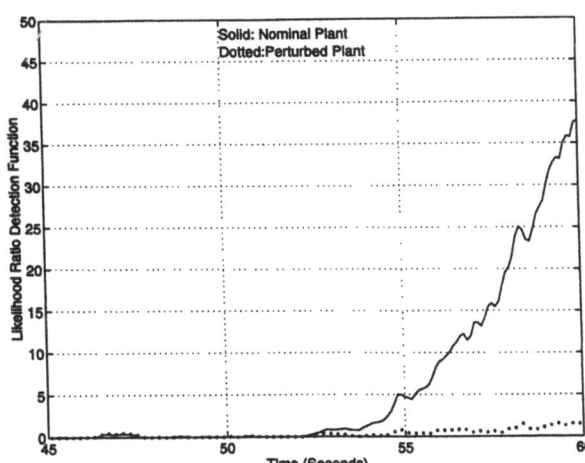

Figure 5.7: The likelihood ratio test's detection function for the nominal and perturbed plants with a ramp failure.

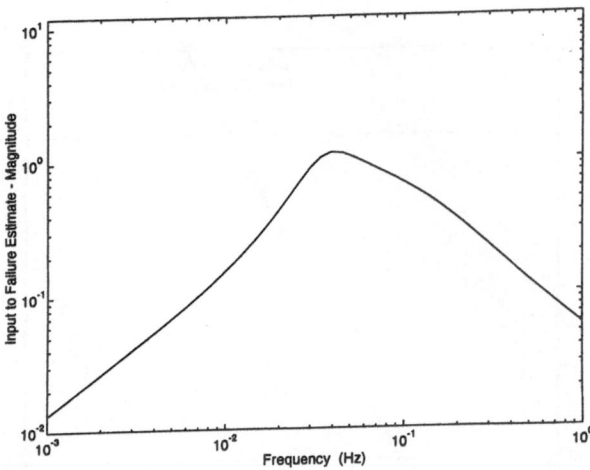

Figure 5.8: Magnitude of frequency response from the stern plane command to the failure estimate for the perturbed plant.

in 15 seconds. With the nominal plant, the solid curve indicates that the decision function starts to rise within 5 seconds after the failure's injection. For the perturbed plant, the detection function, shown by the dotted curve, has yet to rise 10 seconds after the failure is injected.

Maneuvers

We assume again that the vehicle has a constant velocity. For this example, we will consider the effect on the likelihood ratio test's detection function of a known input from the pitch command. The test is designed based on a nominal plant, but is used with a perturbed plant as well.

A plot of the magnitude of the transfer function between the input u and $\hat{f}^c$ is shown in Figure 5.8 for the perturbed plant. The magnitude of this transfer function can be quite high at frequencies where the model uncertainties are large. In the absence of model uncertainties, the magnitude is 0 for all frequencies.

Figures 5.9 and 5.10 illustrate the effects of additive uncertainties on the detection function of the likelihood ratio test. The top plot

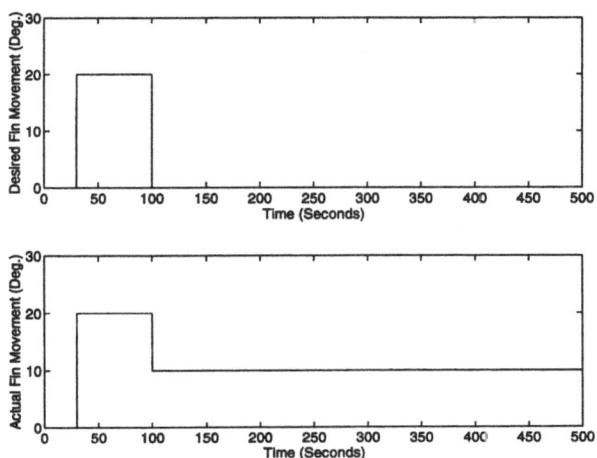

Figure 5.9: Time history of desired (top) and failed (bottom) fin movement.

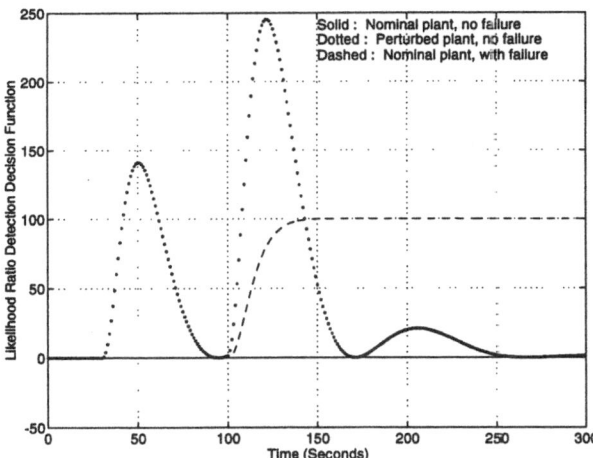

Figure 5.10: The likelihood ratio test's detection function with a maneuver.

of Figures 5.9 shows the desired fin movement for a particular pitch maneuver. At time $t = 30$ seconds, the fin is to be deflected to a 20 degree position, and is to remain there until $t = 100$ seconds. At that time, the fin is commanded to return to the 0 degrees position. The bottom plot shows the actual deflection, if a failure occurs. On the way back to the 0 degree position, the fin gets stuck at 10 degrees. The net effect is a step failure of 10 degrees magnitude injected at time $t = 100$ seconds.

Figure 5.10 shows three plots that represent the behavior of the decision function for three different simulations of the undersea vehicle performing the above pitch maneuver. The solid curve represents the decision function for the nominal plant in the absence of a failure, i.e. when the fin moves as desired. The curve remains at level 0 during the entire maneuver. The dotted curve represents the same decision function for the perturbed plant, also in the absence of failure. With the start of the fin movement at time $t = 30$ seconds, the decision function experiences a sharp rise. A second rise is experienced when the fin is commanded to move again, at time $t = 100$ seconds. Finally, the dashed curve represents the decision function for the nominal plant in the presence of a failure, i.e. when the fin is stuck at 10 degree beginning at time $t = 100$ seconds. The decision function rises as it should.

This example is similar to the one given in Section 2.3.3, and it shows that it is not possible to select a threshold level for some combination of maneuvers and failures. Any choice of threshold level is either too low to avoid a false alarm in the presence of model uncertainty, or too high to allow for the failure to be detected shortly after it occurs in the nominal system. We shall discuss this example again in the next chapter.

5.5   FDI with Robustness to Failure Mode, Noise and Plant Uncertainties

In the last section, we have shown how model uncertainties can seriously degrade the performance of the likelihood ratio test, which assumes accurate plant models. In this section, we describe a robust failure detection and isolation algorithm that is insensitive to failure mode, noise and plant model uncertainties.

Figure 5.11 describes the logic of the algorithm design. The in-

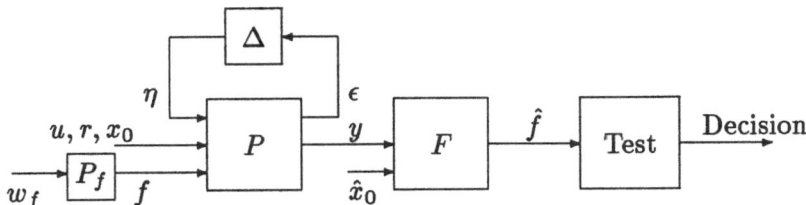

Figure 5.11: Representation of the robust FDI algorithm.

puts to the plant P are the perturbation's output, η, the known inputs, u, the disturbance, r, the initial condition, x_0, and the failure, f, which is assumed to be the output of a shaping filter P_f. As discussed in Section 5.2.3, this failure model embraces a large class of failure modes, making the algorithm robust to failure mode uncertainties. The measurement y is to be fed to the robust estimator, along with the initial state estimate $\hat{x}_0$. The outputs are state and failure estimate that are robust to failure mode, plant and noise modeling errors. The test's detection and isolation functions are based on that estimate.

Finally, Fig. 5.11 is a general description of failure detection and isolation algorithms. For the test described in Chapter 2, P_f is simply a unit step function, $\Delta = 0$, F represents the Kalman filter, and the block 'Test' compares the square of the step magnitude estimate to a threshold. For the likelihood ratio test of Section 5.3, P_f is a shaping filter, while $\Delta = 0$, F is game theoretic filter or smoother, and the block 'Test' represents the comparison of the square of a failure estimate to a threshold. The block 'Test' also uses the failure estimate for isolation.

The decision function is described in Section 5.5.1. The robust estimator is described in Section 5.5.2, and a summary of the algorithm is given in Section 5.5.3.

5.5.1   The Decision Function

In Section 5.3, we have shown that the likelihood ratio test is based on a quadratic function of the observation signal. This decision function can also be described as a function of a failure estimate obtained by minimizing the energy in the failure estimation error. That is, the detection function of Eq.(5.33), rewritten below, is

$$\mathcal{D}^s = \hat{\underline{f}}^{s'} \Sigma_{\underline{f}|Y}^{-1} \hat{\underline{f}}^s \tag{5.43}$$

where $\hat{f}^s$ is obtained by solving the optimization problem

$$\min_{\hat{f}^s} E\left(\|\underline{f} - \hat{\underline{f}}^s\| \mid Y\right) \tag{5.44}$$

subject to the nominal plant and failure model constraints of Eqs.(5.37-5.40). Again, this smoothed estimate is robust to noise model uncertainty since it is the solution to the H_∞ smoothing problem, as shown in Section 3.5. For the filtered version, the decision function

$$\mathcal{D}^c_{k_0+N} = \sum_{k=k_0}^{k_0+N} \|\hat{f}^c_k\|^2_{\Sigma_k^{-1}} \tag{5.45}$$

is obtained by minimizing

$$\min_{\hat{f}} \gamma^2 \log\, E\left(e^{\gamma^{-2}J}\right) \tag{5.46}$$

where J is given by Eqs.(5.35-5.36) (Note we have used $\gamma^{-2} = \theta$), subject to the same constraints.

The game theoretic smoother and filter are, respectively, the solutions to the above two optimization problems. Again as $\gamma \to \infty$, we have the traditional Kalman filter based test.

In the context of our algorithm, an exact model is not assumed. Instead, a bound on the induced 2-norm of the plant perturbation can be assumed, and likewise for the ℓ_2 norm of the disturbance over an interval. A reasonable detection function is one which, in the absence of noise and plant model uncertainties, can reduce to that of the likelihood ratio test, i.e. Eq.(5.33) or (5.41). A smoothed version would be

$$\mathcal{D}^{rs}_{k_0+N} = \sum_{k=k_0}^{k_0+N} \hat{f}^{rs'} S_k \hat{f}^{rs} \tag{5.47}$$

where $\hat{f}^{rs} = \hat{f}^{rs}(Y)$, the robust smoothed estimate, minimizes the worst case error energy, or ℓ_2 norm

$$\min_{\hat{f}^{rs}(Y)} \max_{r,\vartheta,\Delta,\hat{x}_0} \sum_{k=k_0}^{k_0+N} \left(\|f_k - \hat{f}^{rs}_k(Y)\|^2_{S_k}\right) \tag{5.48}$$

subject to the plant and failure models, as well as the constraints on the noise and the perturbations, rewritten below

$$\|u\|^2 + \|r\|^2 + \|\vartheta\|^2 + \|x_0 - \hat{x}_0\|^2_{P_0^{-1}} < 1 \qquad (5.49)$$

$$\|\Delta\|_{i2} < 1 \qquad (5.50)$$

Again, thanks to rescaling, there is no loss of generality implied when using a unity bound. The argument Y in Eq.(5.48) is added in order to emphasize the fact that, at each time step k, $\hat{f}_k^{rs}$ is a function of the entire set of observations on $[k_0, k_0 + N]$. The elements of the matrix $\mathcal{S}_k$ are free parameters. One possibility is to choose a constant diagonal weighting matrix $\mathcal{S}$, where each entry is the reciprocal of the ℓ_2 norm bound on the failure $f_{i,.}$ (Eq.5.2).

The filtered version of Eq.(5.47) is

$$
\begin{aligned}
\mathcal{D}_{k_0+N} &\equiv \mathcal{D}^{rc}_{k_0+N} \\
&= \sum_{k=k_0}^{k_0+N} \hat{f}_k^{rc'} \mathcal{S}_k \hat{f}_k^{rc} \\
&= \sum_{k=k_0}^{k_0+N} \hat{f}_k' \mathcal{S}_k \hat{f}_k
\end{aligned}
\qquad (5.51)
$$

subject to the same constraints as the smoothed version. The superscript rc, for robust causal, will be dropped from now on for the sake of simpler notation. The failure estimate is obtained by replacing the objective function (5.48) with

$$\min_{\hat{f}} \ \max_{r,\vartheta,x_0,\Delta} \ \sum_{k=k_0}^{k_0+N} \left(\|f_k - \hat{f}_k\|^2_{\mathcal{S}_k} \right) \qquad (5.52)$$

The above decision function has a stochastic interpretation. As explained in Chapter 4, if we assume that the noise is Gaussian, then $\hat{f}$ is the estimate that minimizes the risk sensitive objective function. For the filtered estimate, we have

$$\hat{f} = \arg\min_{\hat{f}} E\left(e^{\gamma^{-2} \sum_{k=k_0}^{k_0+N} \|f_k - \hat{f}_k\|^2_{\mathcal{S}_k}} \right) \qquad (5.53)$$

subject to plant and failure models, as well as the induced-norm bound constraint on the perturbation of Eq.(5.50). If a stochastic interpretation is assumed, then a constant diagonal matrix $\mathcal{S}$ can be used, where the diagonal entries are equal to the reciprocals of the variances of the failures, based on the Gauss-Markov model of Eq.(5.11). As mentioned at the end of Section 5.4, the best choice of $\mathcal{S}$ is an open research question. Note that if the failure is a scalar, i.e., only one control command or measurement is of interest, then the question of choosing $\mathcal{S}$ does not arise.

It is obvious from the above discussion that, in order to obtain the desired decision function, we must use a robust filter or smoother, which would give us $\hat{f}^{rc}$ or $\hat{f}^{rs}$. The robust filter is derived in Section 3.3 for discrete-time systems, and in Appendix C for continuous-time systems. The robust smoother is derived in Section 3.5 for the discrete-time case. We discuss the filter design for the robust FDI algorithm in Section 5.5.2.

Finally, note that in the absence of model uncertainties and norm bounds, the robust decision functions of the robust FDI algorithm reduce to the corresponding likelihood ratio detection and isolation functions, provided the appropriate weighting matrix is used, i.e., $\mathcal{S}_k = \Sigma_k^{-1}$. This is by virtue of the fact that robust H_∞ or risk sensitive optimization reduces to linear quadratic optimization in the absence of model uncertainties, as shown in earlier chapters. The robust FDI algorithm is therefore an extension of the likelihood ratio test of Section 5.3, just as the robust H_∞ or risk sensitive estimator is an extension of the maximum likelihood, or more accurately the maximum a posteriori estimator, which is given by the fixed-interval smoother.

5.5.2   Robust Estimator Design

The objective function of the robust filter is:

$$\min_{\hat{f}} \max_{\eta, r, u, \vartheta, x_0} \quad \overline{\mathcal{D}}_{k_0+N} + \|\epsilon\|^2$$

$$-\frac{\gamma^2}{2} \left(\|u\|^2 + \|\eta\|^2 + \|r\|^2 + \|\vartheta\|^2 \right.$$

$$\left. + \|x_0\|_{\overline{X}_0}^2 + \|x_0 - \hat{x}_0\|_{P_0^{-1}}^2 \right) \qquad (5.54)$$

where $\overline{\mathcal{D}}_{k_0+N}$ is given by

$$\overline{\mathcal{D}}_{k_0+N} \equiv \sum_{k=k_0}^{k_0+N} \|\varphi_k - \hat{\varphi}_k\|_{C_f'\mathcal{S}C_f}^2 \qquad (5.55)$$

$$= \sum_{k=k_0}^{k_0+N} \|f_k - \hat{f}_k\|_{\mathcal{S}}^2 \tag{5.56}$$

The last equality follows from the fact that $C_f = I$ for our choice of failure model (Eqs. 5.11-5.15), and the norm of r and η are taken over the interval $[k_0, k_0 + N]$. For the robust smoother, the objective function is the same, except that the failure estimate at each time step k is based on the observations over the entire interval. The constraints are:

$$
\begin{bmatrix} \begin{bmatrix} x_{k+1} \\ \varphi_{k+1} \end{bmatrix} \\ \hline \epsilon_k \\ e_k \\ \begin{bmatrix} y_k \\ u_k \end{bmatrix} \end{bmatrix} = \mathcal{A} \begin{bmatrix} \begin{bmatrix} x_k \\ \varphi_k \end{bmatrix} \\ \hline \eta_k \\ r_k \\ \vartheta_k \\ u_k \\ \hat{\varphi}_k \end{bmatrix} \tag{5.57}
$$

where

$$
\mathcal{A} = \begin{bmatrix} \begin{bmatrix} A & FC_f \\ 0 & A_f \end{bmatrix} & \begin{bmatrix} B_\eta \\ 0 \end{bmatrix} & \begin{bmatrix} B & 0 & U \\ 0 & B_f & 0 \end{bmatrix} & 0 \\ \hline \begin{bmatrix} S & 0 \\ S^{\frac{1}{2}}C_f & 0 \end{bmatrix} & T & \begin{bmatrix} 0 & 0 & 0 \\ 0 & 0 & 0 \end{bmatrix} & \begin{bmatrix} 0 \\ -S^{\frac{1}{2}}C_f \end{bmatrix} \\ \begin{bmatrix} C & LC_f \\ 0 & 0 \end{bmatrix} & \begin{bmatrix} R \\ 0 \end{bmatrix} & \begin{bmatrix} D & 0 & W \\ 0 & 0 & I \end{bmatrix} & \begin{bmatrix} 0 \\ 0 \end{bmatrix} \end{bmatrix} \tag{5.58}
$$

If the model of Eq.(5.11) is used, then the elements of A_f and B_f can be chosen so as to obtain rapid and accurate tracking of the failure. As mentioned in Section 5.3, this can be done by choosing the steady-state gain of the transfer function $T_{\hat{f}f}$ between the input failure and the failure estimate to be close to unity for a large bandwidth. That is,

$$| T_{\hat{f}f}(\omega) | \simeq 1 \quad \forall \, \omega < \omega^* \tag{5.59}$$

The frequency ω^* is determined by trial and error, together with the matrices A_f and B_f. The goal is to have a filter that is as fast as possible, without causing a large false alarm probability by having the disturbances pass as a failure or by sacrificing robustness. Note that ω^*, A_f, and B_f need not be the same for the robust and Kalman filter.

The estimates $\hat{f}_k, k \in [k_0, k_0 + N]$ obtained can then be used in Eq.(5.51). The above robust estimator is defined over a finite-time horizon. Therefore, a separate estimator is needed for each time window.

If a steady-state estimator design is used, then only one estimator is required. In this case, $\overline{\mathcal{D}}_{k_0+N}$ in Eq.(5.56) is replaced by

$$\overline{\mathcal{D}} = \sum_{k=0}^{\infty} \|f_k - \hat{f}_k\|_{\mathcal{S}}^2 \qquad (5.60)$$

for a constant matrix $\mathcal{S}$. The detection function of Eq.(5.51) can then be recursively computed

$$\mathcal{D}_{k_0+N} = \mathcal{D}_{k_0+N-1} - \|\hat{f}_{k_0-1}\|_{\mathcal{S}}^2 + \|\hat{f}_{k_0+N}\|_{\mathcal{S}}^2 \qquad (5.61)$$

Finally, the objective $\overline{\mathcal{D}}_{k_0+N}$ is related to the detection decision function. The same robust estimate can also be used for the isolation decision functions and, if a steady-state linear time-invariant filter is employed, these functions can also be computed recursively in a similar way.

With the objective function of Eq.(5.54), the actual weighting matrix on the plant (including the uncertainty shaping filter) and failure states is of the form

$$M = \begin{bmatrix} 0 & C_f \mathcal{S}^{\frac{1}{2}} \end{bmatrix} \qquad (5.62)$$

where the length of the 0 row subvector is equal to the number of non-failure states. In this case, the minmax optimization is only concerned with the failure states. With a more general weighting matrix in the objective function, say

$$M = \begin{bmatrix} M_x & 0 \\ 0 & C_f \mathcal{S}^{\frac{1}{2}} \end{bmatrix} \qquad (5.63)$$

then the plant states and the failure states are both available. Estimates of the failure states can be used to compute the decision functions as before, while estimates of the plant's states can be used in lieu of those given by, say, an estimator based on the no-failure assumption. This is particularly useful as an estimator designed for the no-failure assumption will not give an accurate estimate of the states after a failure occurs. If a separate filter based on the unfailed model is used to estimate the plant states, then one possibility is to use a filter dedicated to the failure states, i.e., with $M_x = 0$, for detection and isolation, and a parallel one with weights on the plant as well as the failure states. Once a failure is detected, then the second filter can be used to supply estimates of the plant states and failure states, at least until the failure is isolated and accommodated. It should be noted that when the failure estimate is used for purposes other than detection or isolation, then a lower bandwidth failure model could be used, since noise rejection becomes as important as rapid filter response.

We have just made the observation that the robust filtering methodology is versatile enough to allow the designer to focus on a subset of the states by using the weighting matrix M in the objective function of the filter. Other methods, which involve the fine tuning of the plant uncertainty matrices Q, R, S, and T, can also be used for the same purpose. This versatility is very powerfull, as it is often not possible to have one filter that achieves all our objectives. But It is usually possible to design several robust filters, with each filter dedicated to one or more plant or failure states, but valid for an entire family of plants. We exploit this flexibility for the reentry vehicle attitude control system application of Section 6.3.

Robust Estimator Equations

We now give the recursive equations of the robust filter just described. Define

$$\check{A} = \begin{bmatrix} A & FC_f \\ 0 & A_f \end{bmatrix} \tag{5.64}$$

$$\check{B} = \begin{bmatrix} B & 0 & U \\ 0 & B_f & 0 \end{bmatrix} \tag{5.65}$$

$$\check{C} = \begin{bmatrix} C & LC_f \\ 0 & 0 \end{bmatrix} \tag{5.66}$$

$$\check{D} = \begin{bmatrix} D & 0 & W \\ 0 & 0 & I \end{bmatrix} \tag{5.67}$$

Then, as derived in Chapter 3,

$$\begin{bmatrix} \hat{x}_{k+1} \\ \hat{\varphi}_{k+1} \end{bmatrix} = \left(\overline{A}_k - K_k \overline{C}_k \right) \begin{bmatrix} \hat{x}_k \\ \hat{\varphi}_k \end{bmatrix} + K_k y_k \tag{5.68}$$

$$K_k = \left(\overline{B}_k \overline{D}'_k + \overline{A}_k H_k \overline{C}'_k \right)$$
$$\left(\overline{D}_k \overline{D}'_k + \overline{C}_k H_k \overline{C}'_k \right)^{-1} \tag{5.69}$$

where

$$\overline{A}_k = \check{A} + \gamma^{-2} \check{B} Z_k^{-1} F'_k \tag{5.70}$$

$$\overline{B}_k = \check{B} Z_k^{-1/2} \tag{5.71}$$

$$\overline{C}_k = \check{B} + \gamma^{-2} \check{D} Z_k^{-1} F_k' \tag{5.72}$$

$$\overline{D}_k = \check{D} Z_k^{-1/2} \tag{5.73}$$

$$F_k = S'T + \check{A}' X_{k+1} \check{B} \tag{5.74}$$

$$H_k = \left(P_k^{-1} - \gamma^{-2} M_k' M_k \right)^{-1} \tag{5.75}$$

$$Z_k = I - \gamma^{-2} \left(T'T + \check{B}' X_{k+1} \check{B} \right) \tag{5.76}$$

and the matrices X_k and P_k are, respectively, positive definite solutions to the two Riccati equations:

$$X_k = \check{A}' X_{k+1} \check{A} + S'S + \gamma^{-2} F_k Z_k^{-1} F_k'$$
$$X_{k_0+N} = 0 \tag{5.77}$$

$$P_{k+1} = \left(\overline{A}_k - K_k \overline{C}_k \right) H_k \left(\overline{A}_k - K_k \overline{C}_k \right)'$$
$$+ \left(\overline{B}_k - K_k \overline{D}_k \right) \left(\overline{B}_k - K_k \overline{D}_k \right)'$$
$$P_0 \quad \text{given} \tag{5.78}$$

5.5.3   Summary of the Algorithm

The detection hypotheses are given by Eqs.(5.5-5.6), with a failure model given by Eqs.(5.11-5.15) for each failure. The bandwidth and amplitude of each failure's model can be chosen so as to provide rapid failure tracking, as Eq.(5.59) indicates. A general Gauss-Markov model, such as the one given by Eq.(5.9-5.10), can also be used for failure representation. The test's robust decision function is given by Eqs.(5.47-5.48) or Eqs.(5.51-5.52), depending on whether it is based on a smoothed or filtered estimate, respectively.

The failure estimate is provided by the robust game theoretic or risk sensitive estimator in Section 5.5.2. This estimator is an extension of both the Kalman and H_∞ filter. The estimator is based on an augmented model of the plant and the failure states, as shown by Eqs.(5.57-5.58). While the shaping filter of Eqs.(5.15-5.11) makes the algorithm robust to failure mode uncertainties, the choice of filter makes it insensitilve to noise and plant model uncertainty. We have also shown that the robust FDI algorithm is an extension of likelihood ratio tests used when accurate plant and disturbance models can be assumed.

There are two open issues. The first, mentioned at the end of Section 5.4 and following Eq.(5.53), deals with the selection of the normalization matrix, or weights on the failure estimates, that produces the best performing decision function. The second issue is that of threshold selection. Currently, there are no analytical methods that describe the tradeoff between false alarm rate and detection speed for a particular threshold, even in the absence of plant model uncertainty. One must therefore rely on numerical simulation.

5.6 Summary

In this chapter, we described two failure detection and isolation algorithms. One, based on the likelihood ratio test, is robust to failure mode and noise model uncertainties, but assumes accurate knowledge of the plant dynamics. It relies on the minmax or game theoretic filter or smoother for an estimate of the failure signal. We have also shown that the performance of the likelihood ratio test can degrade considerably in the presence of either noise or plant model uncertainties.

The second algorithm is robust to failure mode as well as plant and disturbance model uncertainties. The robust algorithm is an extension of the likelihood ratio test. That is, in the absence of model uncertainties, and with the proper choice of parameters, it reduces to the likelihood ratio test. If, in addition, the failure mode is fixed, then both the robust algorithm and the likelihood ratio test reduce to the generalized likelihood ratio test of [114].

An interesting feature of the robust FDI algorithm is that the robust filter it uses can be designed so as to provide an estimate of the plant's state immediately after a failure occurs. Finally, the algorithms could have a geometric interpretation, as the failure estimate it provides assumes a direction that is a function of the failed component. This property can also be used for failure isolation.

CHAPTER 6

TWO APPLICATIONS

6.1 Introduction

In this chapter, we discuss two applications. The first one, introduced in Chapter 5, is concerned with failure detection in an underwater vehicle. The second application, which is both a detection and estimation problem, deals with the problem of monitoring the attitude control system of reentry vehicles during the flight transition phase from space back into the atmosphere. We have already discussed reentry vehicles in Chapter 3, where we designed a model based robust inertial navigation filter.

6.2 Application to an Underwater Vehicle

In this section, we apply the robust detection and isolation algorithm to the pitch dynamics of an underwater vehicle. A brief description of the vehicle pitch dynamics was given in Section 5.4.1. Further description is given in Appendix D. The same robust FDI algorithm design is to be used with two different configurations of the underwater vehicle. As mentioned in Section 5.4.1, isolation for this example is simple. The sign of the roll measurement can be used, along with that of the failure estimate, to isolate between the left and right horizontal fins.

The states of the robust estimators are the three plant states representing the longitudinal dynamics, the failure states, and four additional states that are a consequence of the uncertainty model. The robust estimator design is described in Appendix D.

The open-loop numerical experiments of Section 5.4.1 showed that uncertainties can seriously degrade the performance of a nominal likelihood ratio test. In Sections 6.2.1 and 6.2.2, we demonstrate the ability of the robust FDI algorithm to detect failures in the presence of the same uncertainties.

6.2.1   Straight and Level Cruise

We assume that the vehicle is in steady state, cruising at constant velocity, so that $\sum_{k=k_0}^{K} u_k \approx 0$ (the control simply responds to the noise). Figure 6.1 shows the performance of the likelihood ratio test and the robust FDI algorithm for a ramp failure whose slope is 0.33 degree per second, with a maximum fin deflection of 5 degrees. The failure is injected into one of the vehicle's horizontal fins. The figure compares both tests' probability of detection vs. probability of false alarm at an *equivalent* time for both vehicle configurations. Specifically, the comparison is made at the time the value of the pitch angle of each vehicle reaches 1 degree, in the noise free case, due to the injection of the failure. For the nominal configuration, the vehicle pitch angle reaches 1 degree 8 seconds after the injection of the ramp failure. For the perturbed configuration, the corresponding time is 14 seconds after the failure is injected. The figure shows that the performance of the robust FDI algorithm with both the nominal (dashed curve) and perturbed plants (dash-dotted curve) is comparable to that of the likelihood ratio test's performance for the nominal plant (solid curve). The likelihood ratio test, however, shows very poor performance with the perturbed plant (dotted curve).

Figure 6.2 shows the time response of the two algorithms' detection functions with a *ramp* failure injected at time $t = 50$ seconds: a fin is deflected from 0 to 5 degrees in 15 seconds. When the likelihood ratio test is used with the nominal plant (solid curve), the figure indicates that the decision function starts to rise within 5 seconds after the failure's injection. For the perturbed plant, the likelihood ratio test detection function's rise is visibly slower (dotted curve). The detection function of the robust FDI algorithm starts to rise within 5 seconds for both the nominal (dashed curve) and the perturbed plants (dash-dotted curve).

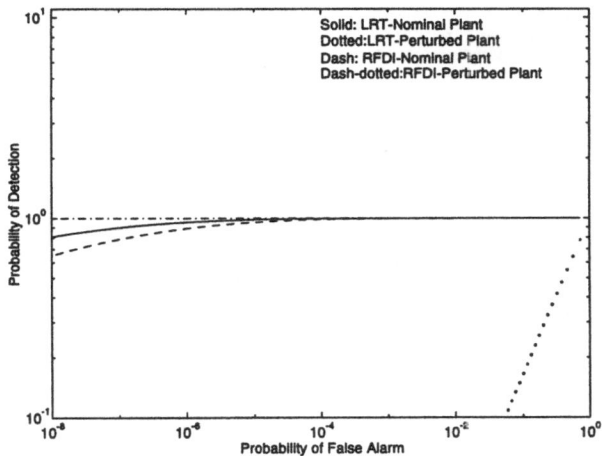

Figure 6.1: Comparison between the robust FDI algorithm and the likelihood ratio test – Probability of detection vs. probability of false alarm with a ramp failure for nominal and perturbed plants.

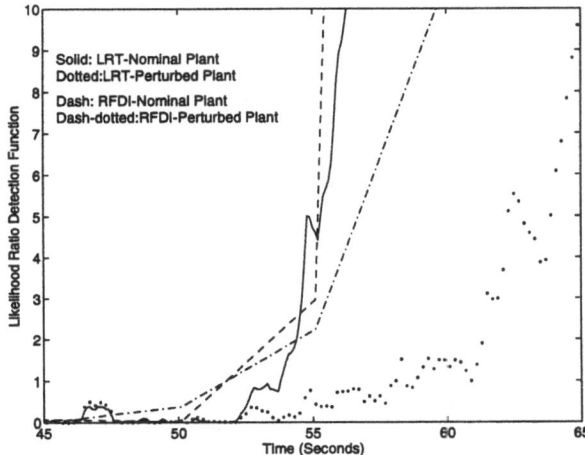

Figure 6.2: The robust FDI algorithm and the likelihood ratio test's decision functions for the nominal and perturbed Plants with a ramp failure.

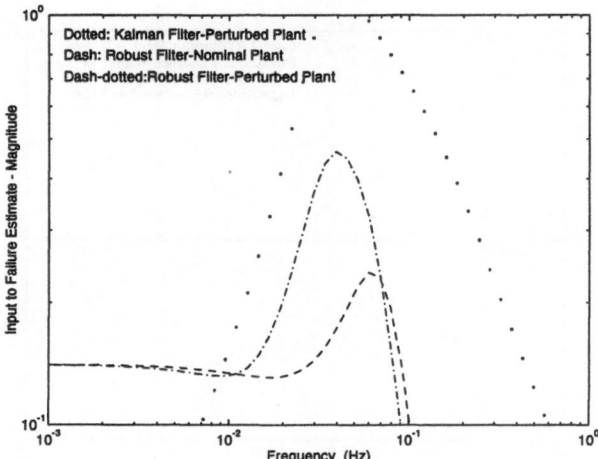

Figure 6.3: Frequency response magnitudes of the transfer functions from the control to the failure estimate for the Kalman filter and the robust estimator.

6.2.2   Maneuvers

We assume again that the vehicle has a constant velocity. For this example, we consider the effect of a known maneuver input from the pitch command on the two algorithms' detection functions, when these algorithms are applied to the two underwater vehicle configurations.

The Bode plot magnitudes of the transfer functions between the input and the failure estimates for the Kalman filter and the robust estimator are shown in Figure 6.3. In the absence of model uncertainties, the magnitude is 0 for all frequencies when the Kalman filter is used with the nominal plant. With the perturbed system, the magnitude of the transfer function when the Kalman filter is used can be quiet high at some frequencies, as shown by the dotted curve. The plots for the nominal and perturbed plant when the robust estimator is used are closer to each other. The projection of the known input onto the robust failure estimate, however, is not insignificant, although significantly smaller than the projection onto the Kalman filter's estimate for the perturbed plant.

Figure 6.4 shows the magnitude Bode plots for the transfer func-

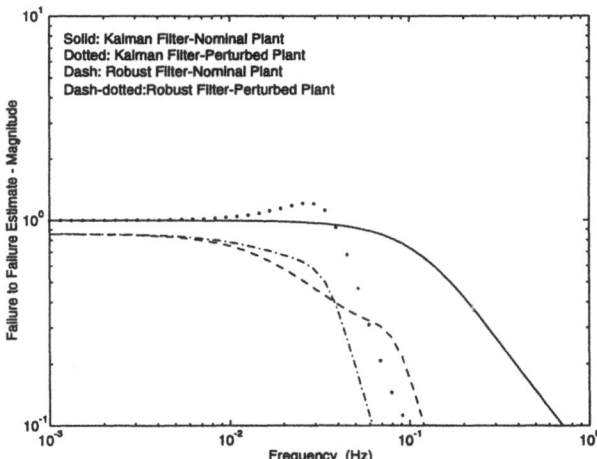

Figure 6.4: Frequency response magnitude of the transfer function from the failure to the failure estimate estimates for the Kalman filter and the robust estimator.

tion from the failure to the failure estimate for both the Kalman filter and the robust estimator. The plots show that: 1) the bandwidth of the Kalman filter is higher than the bandwidth of the robust estimator; 2) the Kalman filter's bandwidth is different for the two plants; and 3) the robust estimators bandwidth is similar for both plants.

The significance of the observations from Figures 6.3 and 6.4 for the time-domain behavior of the detection functions of both tests in the presence of a maneuver is illustrated in Figures 6.6 and 6.7.

The command histories are shown in Figure 6.5. At time $t = 30$ seconds, the fin is commanded to a 20 degree position, and is to remain there until $t = 100$ seconds. At that time, the fin is commanded to return to the 0 degree position, as is shown by the top plot in the figure. The bottom plot shows the actual movement if a failure occurs. On the way back to the 0 degree position, the fin gets stuck at 10 degrees. The net effect is a step failure of 10 degrees magnitude injected at time $t = 100$ seconds.

Figure 6.6 shows the behavior of the likelihood ratio test's detection function in the presence of the maneuver with and without a

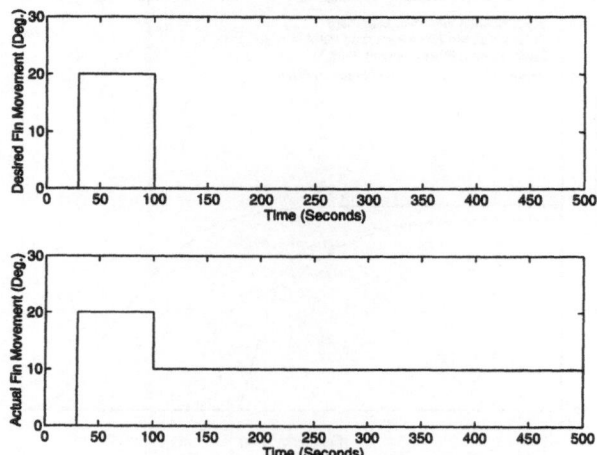

Figure 6.5: Time history of desired (top) and failed (bottom) fin movement.

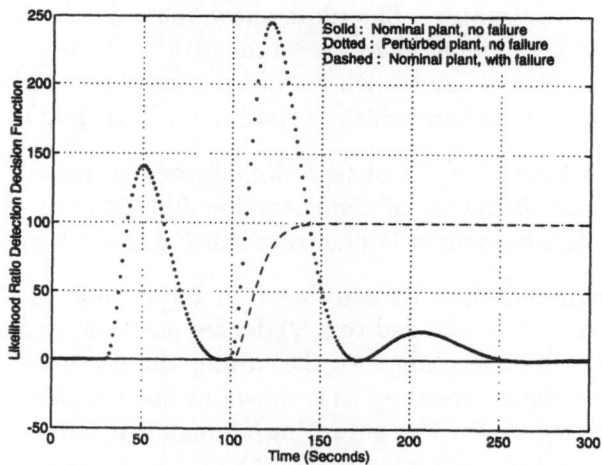

Figure 6.6: The likelihood ratio test's detection function in the presence of a UUV maneuver.

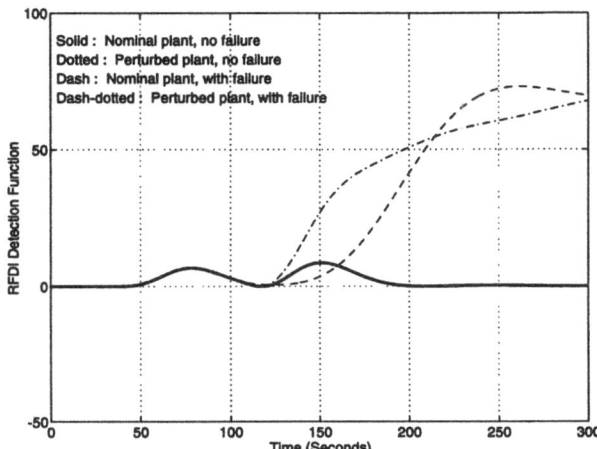

Figure 6.7: The robust FDI algorithm's detection function in the presence of a UUV maneuver.

failure. The likelihood ratio test is disabled by the maneuver: it is not possible to set a threshold. Regardless of the threshold level, either the maneuver will trigger a false alarm with the perturbed model (dotted curve), or the failure will be missed if it occurs with the nominal system (dashed curve).

Figure 6.7 shows the behavior of the robust FDI algorithm's detection function for exactly the same cases. With the robust FDI algorithm, the figure shows that it is possible to set a threshold in such a way that no false alarm is declared because of the maneuver, but the failure is still detected when it occurs in either plant (dashed and dash-dotted curves). A threshold level of, say, 15, would achieve that goal. The figure also shows that a maneuver can cause the robust FDI algorithm's detection function to rise in the absence of failure (the solid and dotted curve, which are coincident), but only very little in comparison with the likelihood ratio test.

In Figure 6.7, if a threshold of 15 is selected, then the failure would be detected within 40 seconds for the one plant, and within 75 seconds for the other. This is much slower, however, than the nominal time to detection with the likelihood ratio test in the absence of model uncertainties, which is less than 20 seconds, as Figure 6.6 shows. This should not be surprising, given the difference in the bandwidths of the

Kalman and robust filters. The results of this example, therefore, show
that robustness could come at a price: it could take longer to detect
a failure. Notice in the previous example, with the vehicle in straight
and level cruise, that no significant price was paid for robustness.

6.3   Application to Reentry Vehicle attitude Control Systems

In this section, we show how robust filters are used in [2] for detecting
and isolating failures in the attitude control system of Reusable Launch
Vehicles (RLV)[1]. In particular, we consider the problem of estimating
the thrust from multiple jets firing from the Reaction Control System
(RCS), as well as the related problem of distinguishing between failures in the RCS and the aerosurfaces. During reentry, plant model
uncertainties are a major problem for the filter as the vehicle's aero-
dynamics vary widely with rapidly changing Mach number, making
gain scheduling impractical. Consequently, the Kalman filter's performance degrades. Even if the Mach number were accurately known,
rapid gain scheduling may not be desirable or even possible, due to
the large data storage requirements it entails. Transient, robust H_∞
or game-theoretic filters are designed for the space shuttle Orbiter's
attitude determination system. Simulation results demonstrate that
the robust filters can be insensitive to plant model uncertainties over
a wide range of Mach numbers, while remaining sensitive to failures in
the aerosurfaces and the RCS jets.

During reentry, Reusable Launch Vehicles (RLV's) typically rely
on aerosurfaces and a Reaction Control System (RCS) for attitude control. For example, the space shuttle Orbiter's aerosurfaces consist of
the rudder, elevons, speedbrake, and body flap, while the RCS consists of bi-propellant jets that are fired in the appropriate direction to
provide desired thrust and augment the aerosurfaces. During on-orbit
operation, only the RCS is used. Shuttle reentry begins with the deor-
bit burn where the Orbiter is oriented in a tail-first position and jets
are fired in order to slow the vehicle and allow capture by the earth's
gravity and atmosphere. The RCS jets reorient the Orbiter to a high
angle of attack, nose-first position. During the first part of reentry, the
Orbiter is oriented to a 40° angle of attack that is maintained until
the vehicle descends and decelerates to below Mach 10. After pass-

---

[1]The results in this section are courtesy of Ramses Agustin.

ing through Mach 10, the Orbiter gradually reduces its angle of attack to 10°. The RCS are the sole attitude effectors until the atmospheric dynamic pressure is large enough for the aerosurfaces to become useful. At some altitude inside the transition region, the control surfaces are activated and attitude control is provided by both the RCS and the aerosurfaces. During the late stages of reentry, control is achieved using the aerosurfaces alone.

Our objective is to detect and isolate failures in the attitude control system. During reentry, the challenge is to distinguish between failures in the aerosurfaces (more specifically, the ailerons, as rudders are not used in the early stages of reentry), and the RCS jets. In addition, we also would like to have an accurate estimate of the jet thrust both in the presence and in the absence of aileron failures. Correct isolation of a failure in the attitude control system is essential to obtaining accurate thrust estimates, which are used for monitoring the health of the RCS. We choose the reentry flight phase because it is the most difficult stage for the attitude control system's failure detection and isolation. Specifically, during reentry, the vehicle's environment undergoes rapid changes in aerodynamic flight properties as the Mach number decreases, and precise knowledge of these flight properties is not available. The combination of rapid change and flight property model uncertainty also makes it difficult to rely on a single, accurate Orbiter model and a Kalman filter design based on that model, or on multiple models and gain scheduling. Moreover, the uncertainty also makes it difficult to distinguish between failures in the RCS and in the elevon control surfaces. Even in the absence of any failure, the Kalman filter's jet thrust estimation performance degrades considerably in the presence of model uncertainty.

Even if the Mach number were accurately known, rapid gain scheduling may not be desirable or even possible, due to the large data storage requirements it entails. In general, however, modeling errors due to inaccurate knowledge of the Mach number and other factors do exist. As a result, the Kalman filter will not yield accurate thrust estimates. A desirable solution, therefore, is a filter architecture that can rapidly distinguish between anomalies in the RCS and in the elevons. The filter architecture must also provide jet thrust estimates even if a jet misfires, or an aerosurface fails.

In the next section, we formulate the problem, and demonstrate the deleterious effect of model uncertainty on the performance of the Kalman filter, motivating the use of robust filtering. In Section 6.3.2, we describe the robust two-filter FDI architecture. Results are presented in Section 6.3.3, and conclusions in Section 6.3.4.

6.3.1   Problem Description

The RCS consists of forty-four bi-propellant jets that, together with the aerosurfaces, provide attitude control and limited three-axis maneuvering capability. Thirty-eight of the jets are primary jets, each capable of providing 870 lbs. of thrust in vacuum. The jets are the sole attitude effectors for the initial part of reentry since there is insufficient dynamic pressure for the control surfaces to be effective in the thin atmosphere. As the vehicle loses altitude, falling into the increasingly denser atmosphere, dynamic pressure increases. The aerosurfaces are then activated and augment the jets until there is sufficient dynamic pressure to control the attitude with the aerosurfaces alone.

We consider the space shuttle Orbiter's lateral dynamics for bank and sideslip. A linear model of the rigid body reentry rotational dynamics is used as derived by Zacharias [124]. While originally given in continuous time, the dynamics are discretized for use with the Orbiter's digital computing system. The state space representation for the Orbiter, linearized at values of the state and control for a selected operating point, are given by

$$
\begin{aligned}
x_{k+1} &= Ax_k + Bu_k + Gw_k + T\theta_k \\
y_k &= Cx_k + Du_k + Ev_k
\end{aligned}
\tag{6.1}
$$

where x_k is the state of the system at time k, u_k is the aerosurfaces control input, θ_k are the jet inputs, w_k is the process noise, y_k is the measurement (angular body rates) vector, and v_k is the sensor noise. In our case, there are six plant states. The first three are: bank angle, angle of attack, and sideslip angle. The remaining three states are the corresponding rates. The measurements, y_k, consist of two linear combinations of bank and sideslip rates. The angle of attack rate is another possible measurement. As the Orbiter descends during reentry, the A, B, and G matrices change in accordance with changes in Mach number and angle of attack. In addition, a new state is defined for the jet thrust estimate. A simple, scalar, high bandwidth Gauss-Markov model is used to represent a multiple-jet firing, i.e.,

$$
\theta_{k+1} = a_\theta \theta_k + g_\theta \nu_k
\tag{6.2}
$$

The parameters a_θ and g_θ determine the bandwidth and amplitude of the model. The above model can be augmented to the plant model of Eq.(6.1) for filter design.

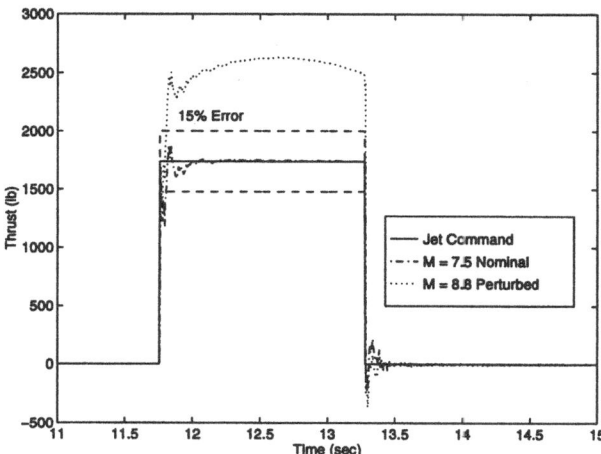

Figure 6.8: Kalman filter's jet thrust estimates for a nominal and perturbed plant.

To motivate the need for robust filtering, a Kalman filter is designed based on the linear, time-invariant state-space model for the nominal condition of $M = 7.5$ and an angle of attack of $\alpha = 35°$. This filter is then tested on both the design plant and on a perturbed plant model based on a Mach number of $M = 8.8$ and $\alpha = 38°$. A step-on-step-off command with two jets firing in the same direction is used in this simple simulation. Aerosurfaces are used for trim adjustment in response to fluctuations in atmospheric conditions as modeled by the process noise.

Simulation results for the Kalman filter are shown in Figure 6.8. The solid line represents the commanded jet thrust magnitude. The dashed lines represent a 15% error margin for the thrust estimate. The jet thrust estimates given by the Kalman filter are represented by the dash-dotted line for the design plant and a dotted line for the perturbed plant.

The estimates for the nominal plant and for the perturbed plant reveal that the Kalman filter works well for the correct plant model, but produces a severely degraded estimate when the model is perturbed.

Note that the error in the degraded estimate for the perturbed plant is
large enough to lead to the conclusion that more jets fired than commanded. As mentioned in the introduction, it is desirable to avoid gain
scheduling due to the rapid change in aerodynamic conditions during
reentry, and the presence of model uncertainty. As is the case with
the attitude determination example of Section 3.6.2, overdesigning the
Kalman filter by increasing the design filter's process noise covariance
did not solve the problem.

In the next section, we describe the filter architecture used for our
robust failure detection and thrust estimation problem.

6.3.2   Robust FDI Filter Architecture

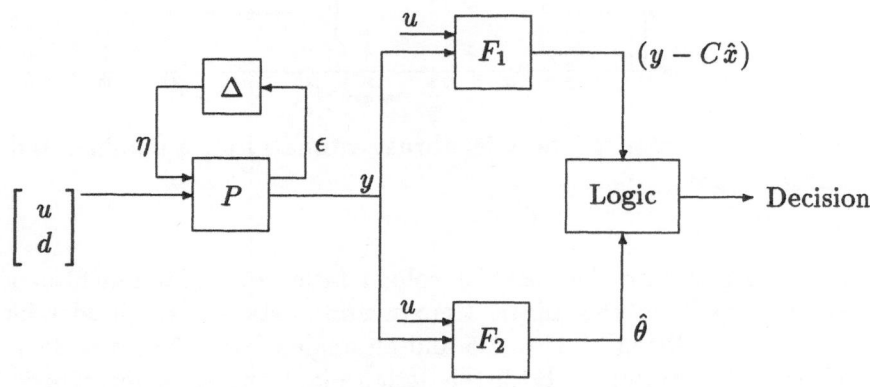

Figure 6.9: Two-filter robust FDI architecture.

Figure 6.9 shows the filter architecture used for attitude control
system fault detection. The output from the plant goes into two robust
filters. As discussed in Chapters 3 and 5, one of the strengths of robust
filter design is that various ways of representing the uncertainty and
weighting matrices are possible. This flexibility can be used to tune in
each of the two filters to a different set of states. This can be done, for
example, by varying the values of the weighting matrix M_k in the filter
equations. Other techniques are also possible, as explained in [2].

As mentioned earlier, the objective is to robustly detect and isolate failures in the lateral dynamics attitude control system, i.e., to

determine whether a failure is in the elevons or in the RCS. An additional goal is to obtain a robust estimate of the jet thrust even if a jet misfires or a failure in the aerosurfaces occurs. Notice that correct isolation is necessary for accurate jet thrust estimation.

In the architecture of Figure 6.9, filter F_1 provides accurate thrust estimates, and can be easily modified should a failure in the aerosurfaces occur. Specifically, the filter, $F1$ is designed so as to robustly minimize the error $\left(\theta_k - \hat{\theta}_k\right)^2$ (Eq.6.2). The design contains 5 states: The bank and sideslip angles, their rates, and the thrust state. The angle of attack states were not needed for this design.

The filter is designed based on a time-invariant plant model, obtained by linearizing around a certain operating point. The goal is to achieve robust performance over a flight envelope, or a range of operating points. This makes gain scheduling at each time step unnecessary, and provides robustness to uncertainties.

The second robust filter, F_2, is the attitude determination filter of Section 3.6.2. It can be utilized to detect failures in the aerosurfaces only, and is therefore insensitive to perturbations in the jet thrust input. This filter essentially keeps the output residual error's norm $\|y_k - \hat{y}_k\|$ low. An alternative design, where an aileron failure state is introduced using a Gauss-Markov model, is also possible. In that case, the objective would be to minimize that state's estimation error. Usually, adding a failure state is necessary to insure sensitivity to the failure. In this approach, however, working with the attitude determination filter's residual was sufficient.

The detection logic works as follows. Once an aileron failure is detected by the second filter, the first filter can be easily modified to provide accurate thrust estimates. Using these two filters in parallel allows for the proper isolation of both kinds of failures in the space shuttle Orbiter's attitude control system, and permits accurate thrust estimates even in the presence of aerosurface failures. Numerical values for the robust filter design may be found in Section D.2. Finally, a one-filter architecture is possible, but the detection response was slower.

6.3.3   Results

The robust FDI architecture just described is tested using the same simulation used to present the Kalman filter results in Section 6.3.1. Nominal and perturbed systems were defined based on different flight regimes. Specifically, parametric perturbations are derived as described in Chapter 3 from two sets of linear time-invariant, system matrices for different flight operating conditions, namely, $M = 7.5, \alpha = 35°$ and $M = 8.8, \alpha = 38°$, where M is the Mach number and α is the angle of attack. Certain nonlinear effects such as jet self-impingement, variation in atmospheric pressure, and other transient effects are ignored. The operating points $M = 7.5$ and $M = 8.8$ were chosen because of the high degree of jet activity present between these two Mach numbers during reentry.

Referring back to Eqs.(6.1), the process noise input matrix G_k is a diagonal matrix whose entries are given by $G_k(i, i) = .05B(i)$ for all rows i to indicate a $\pm 5\%$ uncertainty in aerosurface deflections. A sampling interval of T = 0.005 seconds is chosen to achieve the minimum average estimation error in shortest convergence time.

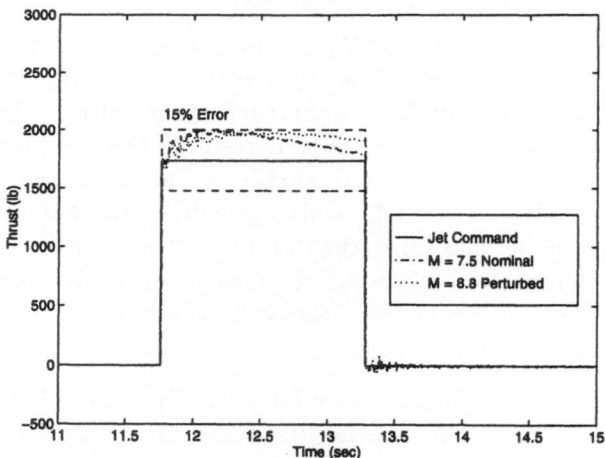

Figure 6.10: Robust filter's (F_1) jet thrust estimates for a nominal and perturbed plant.

We will first discuss results for jet thrust estimation, which is given by filter F_1 in the architecture of Figure 6.9. Simulation results comparing the performance of the Kalman and robust filters are shown in Figure 6.8 (Section 6.3.1) and Figure 6.10. The filters are restarted at the start and end of the jet firing sequence, i.e., whenever the number of jets commanded changes. These plots show the effectiveness of the robust filter's insensitivity to model uncertainty. The dashed lines indicate the 15% error margin, and it is clear that the robust filter estimates are within that margin for both nominal and perturbed plants, while the Kalman filter estimates lie outside the error margin for the perturbed plant. Note also that the performance of the robust filter in Figure 6.10 for either the nominal or perturbed plant compares well with that of the nominal, or optimal, Kalman filter shown in Figure 6.8.

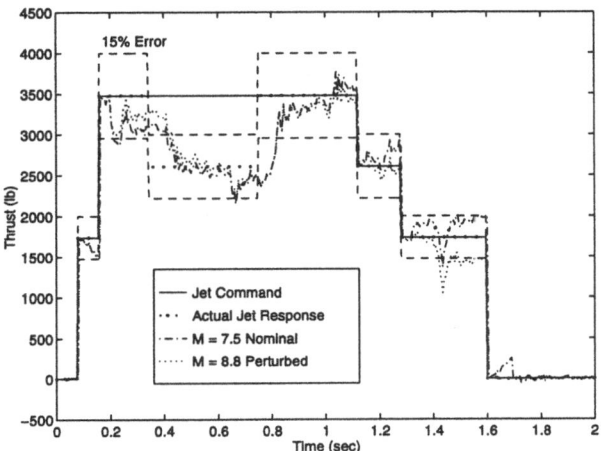

Figure 6.11: Jet failure detection using a transient, robust filter (F_1).

In this simulation, jets were fired during three intervals over a fifteen second simulation period. Figure 6.11 shows the robust filter's transient performance during the first two seconds of that simulation period. One of four jets commanded fails to fire at 0.35 seconds, and remains off until 0.74 seconds. The spikes in the plots of Figure 6.11 are due to the filter's restarting. Overall, these results demonstrate clearly that while the transient robust filter is insensitive to model perturbation, it is highly sensitive to unexpected jet malfunction.

Results comparing the Kalman filter with the Robust filter over the entire fifteen-second simulation period are presented in Table 6.1. Two different simulation runs are used, one where jets fire as commanded, and the other with jet failures. The firing patterns commanded are the same for both simulations, and are shown in part in Figures 6.8 to 6.11. The sum of squared error is the performance measure used for comparison. The term unfailed jet in the table refers to a pattern without jet failure, while the term failed jet refers to a firing pattern with jet failures, exactly as the one shown in Figure 6.11.

The steady-state filters are reset only when new jets are fired while the transient filters are reset periodically.

	Kalman filter	*Robust filter*
Unfailed jet (steady-state)		
Nominal Plant ($M = 7.5$)	1.9282	2.3329
Perturbed Plant ($M = 8.8$)	22.990	3.8438
Unfailed jet (transient filter)		
Nominal Plant	5.0161	2.8658
Perturbed Plant	23.718	5.0349
Failed jet (steady-state)		
Perturbed Plant	6.4095	5.5288
Perturbed Plant	23.655	7.3224
Failed jet (transient filter)		
Perturbed Plant	6.8391	4.4167
Perturbed Plant	21.799	6.7050

Table 6.1: Sum of squared error (10^7 lb^2) in jet estimate.

For all cases shown in the table, whether jets fail or not, and whether a steady-state or transient filter is used, the Kalman filter estimates are severely degraded in the presence of plant perturbation. By comparison, the robust filter gives similar nominal and robust performance. It is also clear that what the robust filter gives up in nominal performance when compared to the Kalman filter, it more than recovers in robust performance.

Moreover, the transient robust filter outperforms the transient Kalman filter, even for the nominal plant. This is because the high bandwidth Gauss-Markov model of Eq.(6.2) used to represent jet thrusts is only approximate, as no linear model can represent a step input. The table also shows that for failure tracking, a transient filter is preferable. When a failure does not occur, however, then the steady-state filter gives slightly better performance. This is because the transient filter is restarted periodically, trading off two different objectives: minimization of error throughout the simulation and speed of detection.

We now discuss the performance of the robust FDI architecture for failure isolation. The objective here is to distinguish between failures in the RCS and elevon control surfaces using both filters in the

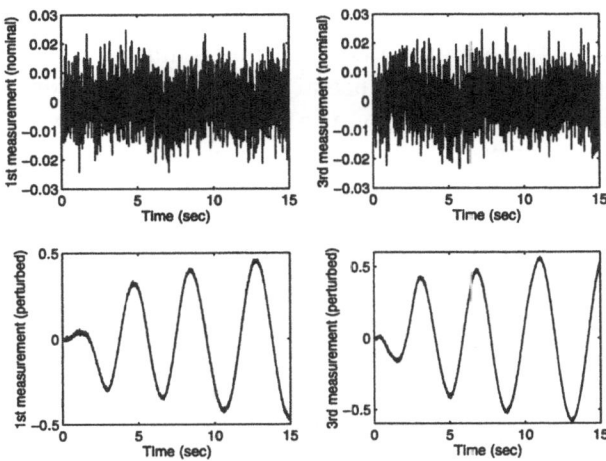

Figure 6.12: Output residuals for the Kalman filter in the absence of failures for the nominal plant (above) and perturbed plant (below).

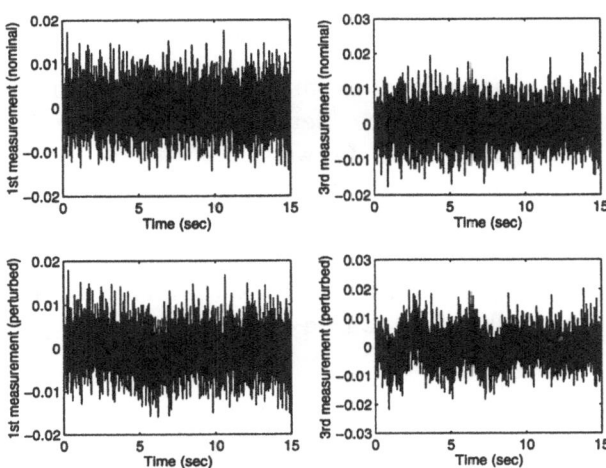

Figure 6.13: Output residuals for the robust filter F_2 in the absence of failures for the nominal plant (above) and perturbed plant (below).

architecture according to the decision logic described in the previous section. Figure 6.12 shows the output residuals for a Kalman filter designed to detect failures in the ailerons. The jet firing pattern is the same as that of the nominal fifteen-second simulation time. No failure occurs, but elevons are put to use at approximately 6 seconds. Results for the nominal Kalman filter on the nominal plant show that the residuals remain unchanged, i.e., the filter recognized the elevon command. However, the same filter, when used with the perturbed plant, shows a distinct shift in its residual, pointing to a failure that did not occur: a false alarm! Model mismatch and an actual elevon failure may therefore be difficult for the Kalman filter to distinguish in the presence of model uncertainty. Raising the threshold is not a satisfactory solution to the problem of avoiding false alarms, since it will result in missed detections.

The robust filter residuals, on the other hand, remain essentially unchanged for both the nominal and perturbed plants, as shown in Figure 6.13. This robust filter design is therefore insensitive to model uncertainty.

Figure 6.14 shows the results in the presence of failure when using the robust filter design. The elevon is failed on (stuck) for 3.5 seconds during the simulation period. The robust filter shows similar shifts in its residuals for both the nominal and perturbed plants. The figure also shows that the same low threshold may be set for both plants, thus keeping the rate of false alarm low.

Figure 6.15 shows the jet estimate when an elevon fails on immediately after a jet failure occurs. Between 0.3 seconds and 0.7 seconds, one of four jets fails to fire. Between 0.9 seconds and 1.2 seconds, an elevon remains failed on at one degree deflection. The shift in the residual of the robust filter concerned with the ailerons, F_2, allows for aileron failure isolation. Subsequently, the filter concerned with jet thrust estimation, F_1, is modified so as to compensate for any sensitivity to the elevon failure. As can be seen, adequate jet performance is maintained.

6.3.4   Summary

This application is concerned with the problem of failure detection and isolation for reentry vehicles' attitude control systems. The objectives are to rapidly detect a failure, and to accurately determine whether it occured in the jet Reaction Control System or the aerosurfaces. Cor-

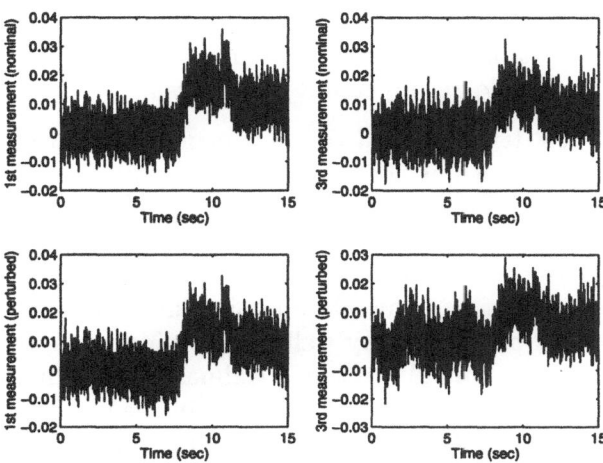

Figure 6.14: Output residuals for the robust filter F_2 after an eleven failure is detected by the same filter.

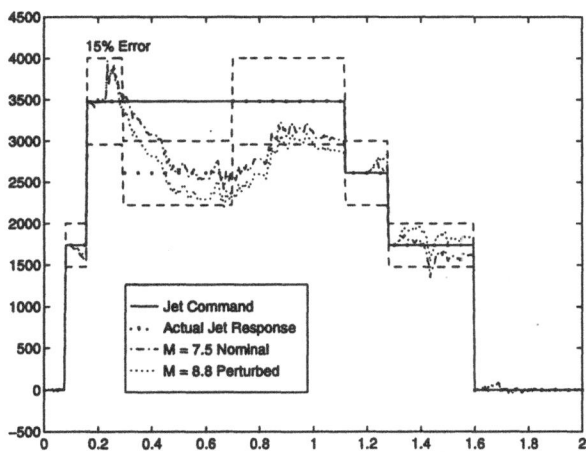

Figure 6.15: Jet failure detection using filter F_1 in the presence of elevon failure.

rect isolation is essential to the second objective, namely, obtaining an accurate estimate of the thrust provided by the jets at all times. The problem is particularly difficult during reentry into the atmosphere, due to both the rapid variations in Mach number, and the large uncertainties in the vehicle aerodynamic properties.

It has been demonstrated that achieving the above objectives requires filtering algorithms that are, on the one hand, robust to model perturbations, and on the other hand, very sensitive to failures in either the jet firings or the aerosurfaces. A Robust failure detection and Isolation architecture that relies on game theoretic/H_∞ filters has been suggested. Implementing this filter in simulation has shown that the robust algorithm performs very well over a wide range of model variations, while the nominally optimal Kalman filter can produce false alarms, and leave some failures undetected, in the presence of the same variations. Further work would include looking in detail at the Reaction Control System jet propulsion model to isolate precisely which jet failed.

APPENDIX A
THE KALMAN FILTER

This appendix presents a derivation of the Kalman filter. The problem is formulated in Section A.1. The solution to this problem is a recursive filter that gives, for linear systems, the minimum error variance state estimate as a linear function of the observations. The derivation of the filter, which is formulated as an optimization problem, is given in Sections A.2 and A.3. In Section A.4, we discuss the special case where the disturbance and initial error are Gaussian. The Gaussian assumption endows the estimate with additional, stronger properties. Section A.5 introduces an important stochastic process, the innovation, which provides more insight into minimum variance estimation. In Section A.6 and A.7, we focus on linear time-invariant systems. The behavior of the filter at steady-state is the subject of Section A.6, while Section A.7 introduces the Wiener filter, or the frequency domain equivalent of the Kalman filter. At each time step, the Kalman filter gives an estimate that is a function of either past (prediction), or past and present (filtering), observations. Estimates that are a function of past, present, and future observations are called smoothed estimates. This is the subject of Section A.8. In Section A.9, we extend the use of the Kalman filter to nonlinear systems, for which no filter that is both optimal and implementable can at present be derived. We introduce the extended Kalman filter (EKF) in Section A.9, which is a suboptimal filter when either the plant or the observations are nonlinear. Finally, a summary of the Kalman filter equations and additional remarks are found in Section A.10[1].

---

[1] Sections A.1 through A.3, and Section A.10 provide the basics of Kalman filtering. The other sections of this appendix contain more in-depth topics.

A.1   Problem Description

Consider the discrete-time linear system

$$x_{k+1} \;=\; A_k x_k + B_k r_k \qquad\qquad (A.1)$$
$$y_k \;=\; C_k x_k + D_k r_k \qquad\qquad (A.2)$$

where x_k is the state of the system at time k, y_k is the observation, and r_k represents the disturbance or noise input to the system. All these vectors are column vectors of real numbers of appropriate dimension. The vector r_k includes both the process and sensor noise. It is assumed to be uncorrelated in time or white, having zero mean and unit covariance. We also assume that an unbiased initial estimate $\hat{x}_0$ with error covariance P_0 is available. The error in the initial estimate is uncorrelated with the noise process r_k, for all k. Thus, we have

$$E\left(r_k\right) \;=\; 0 \qquad\qquad (A.3)$$
$$E\left(x_0\right) \;=\; \hat{x}_0 \qquad\qquad (A.4)$$

and

$$E\left(\begin{bmatrix} r_k \\ (x_0 - \hat{x}_0) \end{bmatrix} \begin{bmatrix} r_j' & (x_0 - \hat{x}_0)' \end{bmatrix}\right) = \begin{bmatrix} \delta_{kj} I & 0 \\ 0 & P_0 \end{bmatrix} \qquad (A.5)$$

where δ_{kj} is the Kronecker-delta function. The extension to the case where the input has nonzero mean is simple, as will be shown in the next section. Furthermore, no loss of generality results from the unit covariance assumption as the matrices B_k and D_k can be adjusted to give any desired covariance. Finally, if the disturbance r_k is correlated in time, then the framework of Eqs.(A.1-A.2) is also applicable, provided the power spectral density of the noise is known, and satisfies some technical conditions. The case where the noise is colored is discussed in the last section of this appendix.

At each time step k, we are looking for two estimates of the state x_k: the one-step predictor, and the filtered estimate,

- **The one-step predictor,** denoted $\hat{x}_k$, is a *recursive* function of all the observations up to time $k-1$. That is, given the plant dynamics and the current predicted state estimate, i.e., the estimate at time $k-1$ prior to collecting measurement $k-1$, the objective is to obtain the estimate of the state at the next step prior to measurement collection, i.e., the estimate at time k prior to collecting measurement k.

- **The filtered estimate,** denoted $\hat{x}_{k|k}$, is a recursive function of the observations up to time k. That is, given the estimate generated by the one-step predictor, the objective is to incorporate the current measurement and refine, or filter, the state estimate[2].

We now give a mathematical formulation of the problem. Define the state estimation errors as

$$\tilde{x}_k \;\equiv\; x_k - \hat{x}_k \qquad\qquad\qquad\qquad (A.6)$$

$$\tilde{x}_{k|k} \;\equiv\; x_k - \hat{x}_{k|k} \qquad\qquad\qquad\qquad (A.7)$$

We require that both recursive estimates be unbiased, meaning that $E(\tilde{x}_k) = E(\tilde{x}_{k|k}) = 0$, where the expectations are taken with respect to $x_k, y_0, \ldots, y_{k-1}$ and $x_k, y_0, \ldots, y_k$, respectively. Furthermore, the two estimates are to minimize the trace of their corresponding error covariances, given by

$$P_k \;\equiv\; E\left((\tilde{x}_k - E(\tilde{x}_k))(\tilde{x}_k - E(\tilde{x}_k))' \right)$$

$$=\; \int \tilde{x}_k \tilde{x}_k' p(x_k, y_0, \ldots, y_{k-1}) dx_k dy_0 \ldots dy_{k-1} \qquad (A.8)$$

$$P_{k|k} \;\equiv\; E\left(\left(\tilde{x}_{k|k} - E(\tilde{x}_{k|k})\right)\left(\tilde{x}_{k|k} - E(\tilde{x}_{k|k})\right)' \right)$$

$$=\; \int \tilde{x}_{k|k} \tilde{x}_{k|k}' p(x_k, y_0, \ldots, y_{k-1}) dx_k dy_0 \ldots dy_k \qquad (A.9)$$

The search for the above estimates can then be formulated as two constrained optimization problems at each step k. Specifically,

$$\min_{\hat{x}_k} \quad \text{trace}\,(P_{k+1}) \qquad \forall k$$
$$\text{subject to} \quad \text{Eqs.(A.1 - A.5)} \qquad\qquad (A.10)$$

and

$$\min_{\hat{x}_{k|k}} \quad \text{trace}\left(P_{k|k}\right) \qquad \forall k$$
$$\text{subject to} \quad \text{Eqs.(A.1 - A.5)} \qquad\qquad (A.11)$$

During the course of the derivation, we will show that the estimate that minimizes the trace of the error covariance also minimizes each of

---

[2]The one-step predictor is sometimes denoted $\hat{x}_{k|k-1}$ or $\hat{x}_k^-$ in the literature. Likewise, the covariance is denoted $P_{k|k-1}$ or P_k^-. We use $\hat{x}_k$ and P_k to simplify notation. The filtered estimate and covariance can also be denoted $\hat{x}_k^+$ and P_k^+, respectively.

the individual variances. In the next section we derive the recursive equations for the one-step predictor, and in Section A.3 we derive the equations for the filtered estimate. In both of these sections, the density functions of the noise and initial error are not specified beyond the first two moments. We do, however, restrict our search to estimates $\hat{x}_k$ and $\hat{x}_{k|k}$ that are a linear function of the observations. That is, we only seek the *linear* minimum variance estimator. In Section A.4 we assume that the noise is Gaussian, without imposing any restriction on the form of the estimator. In that case, though we do not require linearity, the minimum variance estimator turns out to be a linear function of the observations. In fact, it is exactly the same linear estimator that is derived without the Gaussian assumption.

A.2   The One-Step Predictor

We restate our objective: given the current predicted estimate $\hat{x}_k$, to incorporate both the current measurement y_k and the plant dynamics of Eq.(A.1), in order to obtain the best predicted estimate $\hat{x}_{k+1}$ at the next step prior to incorporating the next measurement.

We shall restrict our search to recursive linear estimators of the form

$$\hat{x}_{k+1} = A_{ek}\hat{x}_k + K_k y_k \qquad (A.12)$$

that are unbiased in the sense that

$$E\left(x_k - \hat{x}_k\right) = 0 \qquad \text{if } E(\tilde{x}_0), E(r_\ell) = 0 \ \ \forall \ell < k \qquad (A.13)$$

Such estimators require that A_{ek} takes the form

$$A_{ek} \equiv A_k - K_k C_k \qquad (A.14)$$

To see this, we use an induction argument. From Eq.(A.4), we have $E(x_0 - \hat{x}_0)$. Consider

$$
\begin{aligned}
E\left(x_{k+1} - \hat{x}_{k+1}\right) &= E\left(A_k x_k + B_k r_k - (A_{ek}\hat{x}_k + K_k y_k)\right) \\
&= E\left((A_k - K_k C_k)\, x_k - A_{ek}\hat{x}_k\right) \\
&\quad + (B_k - K_k D_k)\, E\left(r_k\right) \qquad (A.15)
\end{aligned}
$$

where the last equality is obtained by substituting for y_k from Eq.(A.2). Since $E(r_k) = 0$, the third term in Eq.(A.15) vanishes. Assume that, for an arbitrary k, $E(x_k - \hat{x}_k) = 0$. Then the first two terms in Eq.(A.15) vanish as well, and by induction Eq.(A.13) holds for all k if and only if A_{ek} is given by Eq.(A.14). The estimator now takes the form

$$\hat{x}_{k+1} = (A_k - K_k C_k)\hat{x}_k + K_k y_k \tag{A.16}$$

K_k The state estimation error defined in Eq.(A.6) is itself a dynamic system. That is,

$$
\begin{aligned}
\tilde{x}_{k+1} &= x_{k+1} - \hat{x}_{k+1} \\
&= A_k x_k + B_k r_k - (A_k - K_k C_k)\hat{x}_k - K_k y_k \\
&= (A_k - K_k C_k)\tilde{x}_k + (B_k - K_k D_k) r_k \\
&= \tilde{A}_k \tilde{x}_k + \tilde{B}_k r_k
\end{aligned}
\tag{A.17}
$$

where $\tilde{A}_k$ and $\tilde{B}_k$ are defined by comparison of the last two equations. The covariance P_k also satisfies a recursive relationship. Specifically, if we apply the assumptions on the disturbance stated at the beginning of the previous section (zero-mean, white, etc.), we obtain the following Riccati equation

$$
\begin{aligned}
P_{k+1} &= E\left(\tilde{x}_{k+1}\tilde{x}'_{k+1}\right) \\
&= \tilde{A}_k E\left(\tilde{x}_k \tilde{x}'_k\right)\tilde{A}'_k + \tilde{A}_k E\left(\tilde{x}_k r'_k\right)\tilde{B}'_k \\
&\quad + \tilde{B}_k E\left(\tilde{r}_k x'_k\right)\tilde{A}'_k + \tilde{B}_k E\left(r_k r'_k\right)\tilde{B}'_k \\
&= \tilde{A}_k P_k \tilde{A}'_k + \tilde{B}_k \tilde{B}'_k \\
&= (A_k - K_k C_k) P_k (A_k - K_k C_k)' \\
&\quad + (B_k - K_k D_k)(B_k - K_k D_k)'
\end{aligned}
\tag{A.18}
$$

with initial condition P_0. Furthermore, since P_0 is positive semidefinite, it follows that P_k is also positive semidefinite for all k[3]. Note that we have made use of the fact that r_k and $\tilde{x}_k$ are uncorrelated, since the error dynamics are strictly causal, so that r_k affects the error state starting only at time $k + 1$. The problem now reduces to finding, at each time step k, the gain K_k that minimizes the trace of the covariance P_{k+1}. The optimization problem of Eq.(A.10) is now rewritten as

$$\min_{K_k} \quad \text{trace}(P_{k+1}) \qquad \forall k$$

$$\text{subject to} \quad \text{Eq.(A.18)} \tag{A.19}$$

Notice that since P_{k+1} is convex in K_k in Eq.(A.18), its trace is also convex in the same matrix[4]. Substituting for P_{k+1} from the right-hand-side of Eq.(A.18), differentiating trace(P_{k+1}) with respect to K_k, and

---

[3] A $n \times n$ real, symmetric matrix P is positive semidefinite (definite) if, for all vector $x \neq 0$ in R^n, $x'Px \geq 0$ ($x'Px > 0$).

[4] The trace of a matrix is an increasing convex function of the diagonal elements of the matrix, and an increasing convex function of a convex function is convex.

setting the derivative to zero, we obtain[5]

$$
\begin{aligned}
\frac{\partial}{\partial K_k}\text{trace}\,(P_{k+1}) &= 2\,(A_k - K_kC_k)\,P_k\,(-C_k') + 2\,(B_k - K_kD_k)\,(-D_k') \\
&= 2K_k\,(C_kP_kC_k' + D_kD_k') - 2\,(A_kP_kC_k' + B_kD_k') \\
&= 0
\end{aligned}
$$

which gives immediately

$$
K_k = (A_kP_kC_k' + B_kD_k')\,(C_kP_kC_k' + D_kD_k')^{-1} \tag{A.20}
$$

where $C_kP_kC_k' + D_kD_k'$ is assumed to be invertible. This invertibility assumption is satisfied if the measurement error covariance D_kD_k' is invertible, meaning that all measurements are corrupted by noise. Alternatively, if D_kD_k' is singular, then the invertibility of $C_kP_kC_k' + D_kD_k'$ implies that the sensor outputs must be corrupted either by the process noise via $C_kP_kC_k'$, or by the sensor noise, via the term D_kD_k'.

Now, if $z_k = M_kx_k$ is the output whose optimal estimate we seek, then the error covariance to be minimized is $M_kP_kM_k'$. It is easy to see that the optimal gain is still K_k as given by Eq.(A.20). Thus, the Kalman filter gives the linear minimum error variance estimate of any linear combination of the states. It therefore follows, as we mentioned in the previous section, that we have the linear minimum error variance estimate of all states.

Notice that Eq.(A.20), the Riccati equation (A.18) for the covariance is deterministic, even though the plant and the estimator equations are stochastic. It also does not depend on the sample path of the disturbance, but only on the disturbance statistics, and can therefore be computed off line for the entire interval of interest.

If the system experiences a known deterministic input, then the dynamics can be given by

$$
x_{k+1} = A_kx_k + B_kr_k + U_ku_k
$$

where u_k is the deterministic input. The same signal U_ku_k is simply added to the estimator equation, so that

$$
\hat{x}_{k+1} = (A_k - K_kC_k)\,\hat{x}_k + K_ky_k + U_ku_k \tag{A.21}
$$

Since u_k is deterministic, the error covariance equation does not change. Also, if the noise vector r_k has mean m_k, then the one-step predictor

---

[5]For X symmetric, $\frac{\partial}{\partial A}\text{trace}\,(AXA') = 2AX$.

can take the form

$$\hat{x}_{k+1} = (A_k - K_k C_k)\,\hat{x}_k + K_k\,(y_k - D_k m_k) + B_k m_k \qquad \text{(A.22)}$$

Finally, we note that the estimate $\hat{x}_{k+1}$ is the *linear* minimum variance estimate, meaning that it is the best, in the minimum variance sense, only among all estimators of the form of Eq.(A.12).

A.3   Measurement Update and the Filtered Estimate

We now would like to derive the linear *filtered* estimate $\hat{x}_{k|k}$, or the linear minimum variance state estimate that depends on the measurements up to time k, and its associated error covariance $P_{k|k}$. In other words, given the current predicted state estimate, the objective is to obtain a refined or filtered estimate, by incorporating the measurement without projecting to the next time step.

We shall obtain expressions for these two moments in terms of the two moments derived in the previous section, namely, $\hat{x}_k$ and $P_k{}^6$. The solution to this problem is simple if we notice that measurement update is a special case of one-step prediction. Introduce a fictitious step, $k \mid k$, between times k and times $k + 1$, when the plant dynamics are frozen, so that $A_k \equiv I$, and there is no process noise, so that $B_k \equiv 0$. We then have, from Eqs.(A.16, A.18, and A.20),

$$\hat{x}_{k|k} = \hat{x}_k + L_k\,(y_k - C_k \hat{x}_k) \qquad \text{(A.23)}$$

$$P_{k|k} = (I - L_k C_k)\,P_k\,(I - L_k C_k) + L_k D_k D_k' L_k' \qquad \text{(A.24)}$$

$$L_k \equiv P_k C_k'\,(C_k P_k C_k' + D_k D_k')^{-1} \qquad \text{(A.25)}$$

The filtered estimation error, $\tilde{x}_{k|k}$, can be expressed in terms of the prediction error $\tilde{x}_k$. Specifically,

$$\tilde{x}_{k|k} = (I - L_k C_k)\tilde{x}_k - L_k D_k r_k \qquad \text{(A.26)}$$

---

[6]In the previous section, we chose to derive the prediction moments $\hat{x}_k$ and P_k in terms of the prediction moments at step $k - 1$, namely, $\hat{x}_{k-1}$ and P_{k-1}, instead of deriving them in terms of the update moments $\hat{x}_{k|k}$ and $P_{k|k}$. The reason for this choice is that it is not possible to obtain an expression for P_k in terms of $P_{k-1|k-1}$ alone, unless the measurement and process noise are uncorrelated, i.e., unless $B_k D_k' \equiv 0$. The expression would have to include P_{k-1} as well. An expression for P_k in terms of P_{k-1} alone, however, is both possible, and simpler, to derive.

The matrix L_k is usually called the *open loop* Kalman gain, while the matrix K_k derived in the last section is called the *closed loop* Kalman gain. Notice that, as is the case with K_k for the one-step predicted estimate, L_k is the gain minimizing the error covariance of any output $z_k = M_k x_k$ whose optimal estimate we seek. Now, for the case where the process and measurement noise are uncorrelated, so that $B_k D'_k \equiv 0$, then $K_k = A_k L_k$. Substituting for L_k from the Eq.(A.25) into Eq.(A.24), we obtain the following expression (after tedious manipulations)

$$P_{k|k} = P_k - P_k C'_k \left(C_k P_k C'_k + D_k D'_k \right)^{-1} C_k P_k \qquad (A.27)$$

The inverse of the covariance matrix is called the information matrix. Inverting both sides of the above equations using the matrix inversion lemma[7], and assuming all measurements are noisy so that the measurement covariance $D_k D'_k$ is invertible, we have

$$P_{k|k}^{-1} = P_k^{-1} + C'_k \left(D_k D'_k \right)^{-1} C_k \qquad (A.28)$$

Intuitively, the above equation says that the information available on random variable x_k after incorporating measurement y_k is the sum of the information available before y_k is incorporated, or P_k, and the information from the measurement, given by $C'_k \left(D_k D'_k \right)^{-1} C_k$.

If the process and measurement noise are uncorrelated, so that $B_k D'_k \equiv 0$, it is possible to express the predicted estimate in terms of the filtered one. Specifically,

$$\hat{x}_{k+1} = A_k \hat{x}_{k|k} \qquad (A.29)$$
$$P_{k+1} = A_k P_{k|k} A'_k + B_k B'_k \qquad (A.30)$$

We now take a look at the open loop gain L_k for a scalar system, in order to gain some insight into the Kalman filter. We can write

$$L_k = \frac{1}{C_k} \frac{P_k}{\left(P_k + \frac{D_k^2}{C_k^2} \right)}$$

where P_k is now a variance. A large value of L_k in the measurement update of Eq.(A.23) means that the estimate $\hat{x}_k$ depends more on y_k.

---

[7] If A, B, C, and D are arbitrary matrices of appropriate dimensions, then the matrix inversion lemma is the statement $(A + BCD)^{-1} = A^{-1} - A^{-1}B(DA^{-1}B + C^{-1})^{-1}DA^{-1}$.

The gain L_k is in turn inversely proportional to D_k^2, which represents the measurement noise variance. The smaller this term is relative to the prediction error variance P_k, the larger L_k is, and the more we trust y_k. At the limit, if the noise is absent, then $D_k \to 0$, and $L_k \to 1/C_k$. As a result, the measurement update equation reduces to $\hat{x}_{k|k} = (1/C_k)y_k$, implying that no filtering is needed for a state that is perfectly measured. Conversely, the larger the measurement noise variance D_k^2 relative to the prediction error variance P_k, the smaller is the gain L_k, and the greater is the reliance on past measurements, or $\hat{x}_k$. In fact, as $D_k \to \infty$, $L_k \to 0$, so that $\hat{x}_{k|k} = \hat{x}_k$, meaning that the measurement y_k is of no value!

We recall one more time that no assumption was made on the density function of the noise. For this reason, the state estimates $\hat{x}_k$ and $\hat{x}_{k|k}$ obtained thus far are the linear minimum variance estimates, meaning that we restricted our search to linear recursive estimators. The situation is different when the noise is Gaussian, as will be seen in the next section.

A.4   Gaussian Disturbance

In deriving the Kalman filter, we have assumed a linear recursive form for the estimator. In this section, we discard that assumptions, and assume instead that the initial state estimation error $\tilde{x}_0$ is Gaussian, and the r_k's are not only uncorrelated, but also Gaussian. The disturbances (and initial error) are therefore independent of each others for all k[8]. As a result, it is easy to show that the x_k's are Markov, meaning that $p(x_{k+1} \mid x_k, \ldots, x_0) = p(x_{k+1} \mid x_k)$. They are also Gaussian. The x_k's are therefore called a *Gauss Markov* process.

In this section, we show that for Gauss Markov processes whose dynamics and observations are given by Eqs.(A.1-A.2), the linear form of the optimal estimator is a consequence of the Gaussian assumption, and needs not be assumed a priori. We also discuss the additional properties that the Kalman filter estimates acquire when the Gaussian assumption holds.

We now take $\hat{x}_k$ and $\hat{x}_{k|k}$ to be the means of the Gaussian conditional density functions $p(x_k|y_0, \ldots, y_{k-1})$ and $p(x_k|y_0, \ldots, y_k)$, respec-

---

[8]Recall that in general, if two random variables α and β are statistically independent, meaning that $p(\alpha, \beta) = p(\alpha)p(\beta)$, then they are uncorrelated as well. The converse, however, is not true, unless α and β are Gaussian.

tively. Specifically,

$$\hat{x}_k = E(x_k|y_0, ..., y_{k-1}) \tag{A.31}$$
$$\hat{x}_{k|k} = E(x_k|y_0, ..., y_k) \tag{A.32}$$

Our objective is to show that since the noise and initial error are Gaussian, the moments of the conditional density functions of the state satisfy the linear recursive relations given by Eq.(A.16) and Eq.(A.23), respectively. Since the mean of a random variable is its minimum variance estimate[9], it follows that the linear form of the Kalman filter is a consequence of the Gaussian assumption. To derive the estimator, we will make use of the following theorem.

Theorem A.1 *Let α and β be jointly Gaussian random vectors, with respective means $\bar{\alpha}$ and $\bar{\beta}$, respective covariances Σ_α and Σ_β, and joint correlation matrix $\Sigma_{\alpha\beta}$. Then the conditional density function $p(\alpha|\beta)$ is also Gaussian with mean and covariance given by*

$$\hat{\alpha} = \bar{\alpha} + \Sigma_{\alpha\beta}\Sigma_\beta^{-1}(\beta - \bar{\beta}) \tag{A.33}$$
$$\Sigma_{\alpha|\beta} = \Sigma_\alpha - \Sigma_{\alpha\beta}\Sigma_\beta^{-1}\Sigma'_{\alpha\beta} \tag{A.34}$$

Proof [71]: Define $\hat{\zeta} = \bar{\alpha} + \Sigma_{\alpha\beta}\Sigma_\beta^{-1}(\beta - \bar{\beta})$, and $\tilde{\zeta} = \alpha - \hat{\zeta}$. Then $E(\tilde{\zeta}) = 0$, and

$$\begin{aligned}
E\left(\tilde{\zeta}(\beta - \bar{\beta})'\right) &= E\left((\alpha - \bar{\alpha})(\beta - \bar{\beta})'\right) \\
&\quad -\Sigma_{\alpha\beta}\Sigma_\beta^{-1}E\left((\beta - \bar{\beta})(\beta - \bar{\beta})'\right) \\
&= 0
\end{aligned}$$

Thus, since $\tilde{\zeta}$ and β are jointly Gaussian and uncorrelated, they are independent. Hence,

$$\begin{aligned}
\hat{\alpha} &= E(\alpha \mid \beta) \\
&= E(\hat{\zeta} + \tilde{\zeta} \mid \beta) \\
&= \hat{\zeta}
\end{aligned}$$

This proves Eq.(A.33). To prove Eq.(A.34), note that $\hat{\alpha}$ and $\tilde{\alpha} = \alpha - \hat{\alpha}$ are independent[10], which implies that $\Sigma_\alpha = \Sigma_{\hat{\alpha}} + \Sigma_{\tilde{\alpha}}$. But $\Sigma_{\tilde{\alpha}} = \Sigma_{\alpha|\beta}$. Thus, we

---

[9]If x and y are random vectors, with the conditional density of x given y denoted $p(x|y)$, then $z = E(x \mid y)$ is the solution to $\min_z \int ((x-z)'(x-z))\, p(x \mid y)dx$.

[10]The independence of α and $\tilde{\alpha}$ can be shown by writing the appropriate correlation matrices. Alternatively, one can invoke the orthogonality property of minimum variance estimators, which is discussed at the end of this section.

have

$$\begin{aligned}
\Sigma_{\alpha|\beta} &= \Sigma_\alpha - \Sigma_{\hat{\alpha}} \\
&= \Sigma_\alpha - \Sigma_{\alpha\beta}\Sigma_\beta^{-1}\Sigma'_{\alpha\beta}
\end{aligned}$$

◇

We first derive the measurement update step, which gives the filtered estimate. Starting at step $k = 0$, let x_0 and y_0 have joint density $p(x_0, y_0)$. Using Bayes' rule, we have

$$\begin{aligned}
p(x_0|y_0) &= \frac{p(x_0, y_0)}{\int p(x_0, y_0)\, dx_0} \\
&= \frac{p(x_0, y_0)}{p(y_0)}
\end{aligned} \tag{A.35}$$

If x_0 is Gaussian with mean $\hat{x}_0$ and covariance P_0, and r_0 is Gaussian with mean zero and unit covariance, then $y_0 = Cx_0 + Dr_0$ is also Gaussian. In fact, $(x'_0, y'_0)'$ are jointly Gaussian random vectors

$$\begin{bmatrix} x_0 \\ y_0 \end{bmatrix} \sim \mathcal{N}\left(\begin{bmatrix} \hat{x}_0 \\ C_0\hat{x}_0 \end{bmatrix}, \begin{bmatrix} P_0 & P_0C'_0 \\ C_0P_0 & C_0P_0C'_0 + D_0D'_0 \end{bmatrix} \right)$$

where the notation $\alpha \sim \mathcal{N}(\alpha, \Sigma)$ means that random variable α is Normal with mean α and covariance Σ. We would like to obtain the conditional density $p(x_0|y_0)$. From the above theorem, we have

$$p(x_0|y_0) = \mathcal{N}\left(\hat{x}_{0|0}, P_{0|0} \right) \tag{A.36}$$

Theorem A.1 gives the mean $\hat{x}_{0|0}$ and covariance $P_{0|0}$:

$$\begin{aligned}
\hat{x}_{0|0} &= \hat{x}_0 + L_0(y_0 - C_0\hat{x}_0) & \text{(A.37)} \\
P_{0|0} &= P_0 - P_0C'_0(C_0P_0C'_0 + D_0D'_0)^{-1}C_0P_0 & \text{(A.38)}
\end{aligned}$$

where

$$L_0 = P_0C'_0(C_0P_0C'_0 + D_0D'_0)^{-1} \tag{A.39}$$

The above three expressions are none other than Eq.(A.23) and Eq.(A.27), using the gain of Eq.(A.25), all for time-step $k = 0$.

To derive the one-step predicted estimate, we now must obtain the density function of x_1 given the observation y_0, or $p(x_1|y_0)$. Again, since

$$\begin{bmatrix} x_1 \\ y_0 \end{bmatrix} = \begin{bmatrix} A_0 & B_0 \\ C_0 & D_0 \end{bmatrix} \begin{bmatrix} x_0 \\ r_0 \end{bmatrix}$$

it follows that x_1 and y_0 are jointly Gaussian. Specifically

$$\begin{bmatrix} x_1 \\ y_0 \end{bmatrix} \sim \mathcal{N}\left(\begin{bmatrix} A_0\hat{x}_0 \\ C_0\hat{x}_0 \end{bmatrix}, \begin{bmatrix} A_0P_0A_0' + B_0B_0' & A_0P_0C_0' + B_0D_0' \\ C_0P_0A_0' + D_0B_0' & C_0P_0C_0' + D_0D_0' \end{bmatrix}\right)$$

(A.40)

Again, the conditional density $p(x_1|y_0)$ is Gaussian

$$p(x_1|y_0) = \mathcal{N}(\hat{x}_1, P_1)$$

(A.41)

The mean and covariance are obtained again using Theorem A.1. For the mean, we have from Eq.(A.33)

$$\begin{aligned} \hat{x}_1 &= A_0\hat{x}_0 + (A_0P_0C_0' + B_0D_0')(C_0P_0C_0' + D_0D_0')^{-1} \\ &\qquad\qquad\qquad\qquad\qquad\qquad\qquad \times (y_0 - C'\hat{x}_0) \qquad \text{(A.42)} \\ &= (A_0 - K_0C_0)\hat{x}_0 + K_0y_0 \qquad\qquad\qquad \text{(A.43)} \end{aligned}$$

where

$$K_0 = (A_0P_0C_0' + B_0D_0')(C_0P_0C_0' + D_0D_0')^{-1}$$

(A.44)

Applying Eq.(A.34) for the covariance

$$\begin{aligned} P_1 &= A_0PA_0' + B_0B_0' \\ &\quad - (A_0P_0C_0' + B_0D_0')(C_0P_0C_0' + D_0D_0')^{-1}(A_0P_0C_0' + B_0D_0')' \\ &= A_0P_0A_0' + B_0B_0' - K_0(C_0P_0C_0' + D_0D_0')K_0' \\ &= (A_0 - K_0C_0)P_0(A_0 - K_0C_0)' + (B_0 - K_0D_0)(B_0 - K_0D_0)' \\ &\quad + ((A_0P_0C_0' + B_0D_0')K_0' + K_0(A_0P_0C_0' + B_0D_0')) \\ &\quad - 2K_0(C_0P_0C_0' + D_0D_0')K_0' \\ &= (A_0 - K_0C_0)P_0(A_0 - K_0C_0)' + (B_0 - K_0D_0)(B_0 - K_0D_0)' \end{aligned}$$

The last equality follows from the fact that

$$\begin{aligned} K_0(C_0P_0C_0' + D_0D_0')K_0' &= (A_0P_0C_0' + B_0D_0')K_0' \\ &= K_0(A_0P_0C_0' + B_0D_0') \end{aligned}$$

The mean, gain, and covariance given above are simply those of Eq.(A.16), Eq.(A.20), and Eq.(A.18), respectively, for $k = 0$. Note that, earlier, we did not need to derive the filtered estimate directly as it is a special case of one-step prediction (use $A_0 = I$ and $B_0 = 0$).

Using an induction argument, we can apply the above procedure at each step k, after the measurement is taken, and obtain that $p(x_k|y_0, ..., y_k)$, and $p(x_{k+1}|y_0, ..., y_k)$ are Gaussian. The mean $\hat{x}_{k|k}$ and covariance $P_{k|k}$ of $p(x_k|y_0, ..., y_k)$ are given respectively by Eq.(A.23) and Eq.(A.27), using the gain of Eq.(A.25). Likewise, the mean $\hat{x}_{k|k-1}$ and covariance $P_{k|k-1}$ are given respectively by Eq.(A.16) and Eq.(A.18), using the gain of Eq.(A.20).

We can therefore conclude that if the noise is Gaussian, then the Kalman filter equations give the minimum variance estimate of the state, instead of the linear minimum variance estimate, as is the case with non-Gaussian noise. The linear form of the estimator needs not be assumed a priori. It is a consequence of the fact that the disturbance is Gaussian. In the remaining part of this section, we discuss some of the additional properties that the state estimate acquires when the disturbance is Gaussian.

Notice that, if the noise is not Gaussian, then the Kalman filter estimates $\hat{x}_k \neq E(x_k \mid y_0, ..., y_{k-1})$ and $\hat{x}_{k|k} \neq E(x_k \mid y_0, ..., y_{k-1})$, i.e. the estimates are not the conditional expectations of the state. Another difference between the Gaussian and the non-Gaussian case is that, in the Gaussian case, we have

$$
\begin{aligned}
P_k &= E\left(\tilde{x}_k \tilde{x}_k'\right) \\
&= E\left(\tilde{x}_k \tilde{x}_k' \mid y_0, \ldots, y_{k-1}\right) & (A.45) \\
P_{k|k} &= E\left(\tilde{x}_{k|k} \tilde{x}_{k|k}'\right) & (A.46) \\
&= E\left(\tilde{x}_{k|k} \tilde{x}_{k|k}' \mid y_0, \ldots, y_k\right) & (A.47)
\end{aligned}
$$

Thus, in the Gaussian case, where our estimate is the conditional mean, P_k and $P_{k|k}$ are both the unconditioned error covariances and the error covariances conditioned on the observations, while in the nonGaussian case, where we do not have the conditional mean, the same two matrices represent the unconditional error covariances, i.e., the error covariances averaged over all possible observations.

Furthermore, since the mean and covariance uniquely determine a Gaussian density function, the entire conditional distribution is available from the mean and covariance equations in the previous sections. This is not the case if the noise is not Gaussian, as the first two moments do not, in general, uniquely specify the desired density functions. Another observation follows from the fact that the mean of the Gaussian random variable is also its mode, or the most likely value that the density function can take. The state estimates obtained have therefore an additional desirable property: they provide the maximum a posteriori probability estimate of the state, given the observations. Specifically, $\hat{x}_k$ maximizes $p(x_k|y_0, ..., y_{k-1})$, and $\hat{x}_{k|k}$ maximizes $p(x_k|y_0, ..., y_k)$.

Before closing this section, we introduce the **orthogonality** property of minimum variance estimates. If $\hat{x}_k = E(x_k \mid y_0, ..., y_{k-1})$ is the minimum error variance estimate, this means that it is not possible to extract any more information from the observations. The orthogonality property expresses this fact by stating that the observations and the error are uncorrelated[11]. Specifically, if $g(y_0, ..., y_{k-1})$ is any function of the observations up to time $k - 1$, then

$$E\left(g(y_0, ..., y_{k-1})(x_k - E(x_k \mid y_0, ..., y_{k-1}))'\right) = 0 \qquad (A.48)$$

Thus, the observations, or any function thereof, are uncorrelated with the error which, in the Gaussian case, implies that they are statistically independent. To see this, define $Y_{k-1} = (y_0, ..., y_{k-1})$. We then have

$$
\begin{aligned}
E\left(g(Y_{k-1})(x_k - E(x_k \mid Y_{k-1}))'\right) &= E\left(g(Y_{k-1})x_k'\right) \\
&\quad - E\left[g(Y_{k-1})E(x_k' \mid Y_{k-1})\right] \\
&= E\left(g(Y_{k-1})x_k'\right) \\
&\quad - E\left[E\left(g(Y_{k-1})x_k' \mid Y_{k-1}\right)\right] \\
&= E\left(g(Y_{k-1})x_k'\right) - E\left(g(Y_{k-1})x_k'\right) \\
&= 0
\end{aligned}
$$

The first equality follows from the linearity of the expectation operation, the second equality follows from the fact that, given the observations, then any function thereof is deterministic, and the third equation follows from the law of iterated expectations[12].

In the non-Gaussian case, the Kalman filter gives only the linear minimum variance estimate. The orthogonality property still holds, except that the function g is restricted to be a linear function of the observations. In that case, then, the *linear* minimum variance error

---

[11]The name orthogonality derives from a geometric interpretation. In the linear space of zero-mean random variales, one can show that the correlation operation is an inner product.

[12]If α and β are two random variables, then the law of iterated expectations says that $E(E(\alpha|\beta)) = E(\alpha)$, where the first or inside expectation operation is taken with respect to α, and the second or outside expectation is with respect to β. To see this, $E(E(\alpha|\beta)) = \int\int \alpha p(\alpha|\beta)p(\beta)d\alpha d\beta = \int \alpha \left(\int p(\alpha|\beta)p(\beta)d\beta\right) d\alpha = \int \alpha p(\alpha)d\alpha = E(\alpha)$. The law can be used to prove many otherwise difficult results. The challenge in applying the law is to find the right random variable to condition on.

is uncorrelated, but not necessarily statistically independent, of any linear function of the observations.

A.5   The Innovation Process

In this section, we continue to assume that the noise and initial error are Gaussian. It is possible to gain further insight by considering the measurement update equation (A.23) further. This equation can be rewritten

$$\hat{x}_{k|k} = \hat{x}_k + L_k \varrho_k \qquad (A.49)$$

where

$$\varrho_k \equiv y_k - C_k \hat{x}_k \qquad (A.50)$$

is the prediction *residual*. The estimate $\hat{x}_{k|k}$, which is optimal in more than one sense, is a linear combination of the optimal estimate $\hat{x}_k$ based on the measurements through time $k - 1$ and the error term ϱ_k. Consider now $\hat{y}_k$, the estimate of the kth observation given all the past ones. We have

$$
\begin{aligned}
\hat{y}_k &\equiv E\left(y_k | y_0, ..., y_{k-1}\right) \\
&= C_k E\left(x_k | y_0, ..., y_{k-1}\right) + D_k E\left(r_k | y_0, ..., y_{k-1}\right) \\
&= C_k \hat{x}_k
\end{aligned}
$$

so that

$$\varrho_k = y_k - \hat{y}_k \qquad (A.51)$$

or

$$y_k = \hat{y}_k + \varrho_k \qquad (A.52)$$

Thus, $\hat{y}_k$ is the component of the kth measurement that can be predicted using a linear function of the past observations, and ϱ_k represents the measurement prediction error, or the *innovation*, meaning the component carrying the new information that measurement y_k brings.

Since $\hat{y}_k$ is a linear function of $y_0, ..., y_{k-1}$, Eq.(A.51) says that the innovations are a causal linear function of the observations and the initial estimate $\hat{y}_0$. The converse is also true, meaning that the observations are a causal linear function of the innovations. This can be seen from Eq.(A.52). Specifically, for $k = 0$ if $\hat{x}_0$ is given, then $\hat{y}_0$ is the estimate available prior to measurement y_0, and y_0 is simply the sum of that prior estimate and ϱ_0. For $k = 1$, Eq.(A.52) says that y_1 is the sum of ϱ_1 and $\hat{y}_1$, but $\hat{y}_1$ is a function of y_0, which is a function of the innovation ϱ_0 and the initial estimate $\hat{y}_0$. So y_1 can be reconstructed from $\hat{y}_0$, ϱ_0 and ϱ_1. More generally, any y_k can be obtained from $\hat{y}_0$ and $\varrho_1, ..., \varrho_k$. As a result, the mapping between the innovations and the observations is one-to-one, and both processes contain the same statistical information[13].

What makes the innovation process of interest is that the ϱ_k's have a mean of zero and are statistically independent of each other. As a result, each innovation contains information not contained in the other innovations. In Section 2.3.1, the Kalman filter is used in a failure detection algorithm to generate the innovations from the observations. The fact that the innovations are statistically independent allows for an easier implementation of the detection algorithm.

To show that the innovation process has zero mean, we use the law of iterated expectations [88]

$$
\begin{aligned}
E\left(\varrho_k\right) &= E\left(y_k - E\left(y_k | y_0, ..., y_{k-1}\right)\right) \\
&= E\left(y_k\right) - E\left(E\left(y_k | y_0, ..., y_{k-1}\right)\right) \\
&= E\left(y_k\right) - E\left(y_k\right) \\
&= 0
\end{aligned}
$$

That the innovations are independent follows from the orthogonality property.

It is easy to show the following interesting fact about the innovation process. Let x_k and y_k, $k = 0, 1, ...$ be two zero mean, correlated stochastic processes, and let ϱ_k, $k = 0, 1, ...$ be the innovations of the y_k's, then

$$
\begin{aligned}
E(x_k | y_0, ..., y_{k-1}) &= E\left(x_k | \varrho_0\right) + ... + E\left(x_k | \varrho_{k-1}\right) \\
&= \sum_{i=0}^{k-1} E\left(x_k | \varrho_i\right)
\end{aligned} \tag{A.53}
$$

If the processes are not zero mean, then the above formula can be adjusted. We will not do it here, but it is also possible to derive the Kalman filter using the above formula.

Note that both the innovations and the disturbance are white, the first with covariance $\Gamma_k = C_k P_k C_k' + D_k D_k'$, and the second with unit covariance. The innovation can also be standardized by defining $\check{\varrho}_k =$

---

[13] More formally, the observation and innovation processes generate the same σ-algebra.

$\Gamma_k^{-\frac{1}{2}} \varrho_k$, which can be thought of as a standard innovation process[14]. Specifically,

$$\hat{x}_{k+1} = A_k \hat{x}_k + K_k \Gamma_k^{1/2} \check{\varrho}_k \qquad (A.54)$$

We note that in the absence of the Gaussian noise assumption, the ϱ_k's are uncorrelated, but not independent. They can be considered innovations in "the wide sense". We will see in Section A.7 how the whiteness property of the innovation is used to derive the Wiener filter. The reader can find an in-depth study of the innovation process and its use in filtering in a series of articles by Kailath, the first of which deals with the linear case [60].

A.6   Linear Time-Invariant Systems

In this section, we focus our attention on linear time-invariant (LTI) systems, i.e., the case where $A_k \equiv A, B_k \equiv B, C_k \equiv C, D_k \equiv D$. We will state and discuss the more important results without formal derivation. One question of interest is whether the Kalman filter for such systems is also time-invariant. If we apply Eq.(A.18) for $k = 1$ to an LTI system, then P_1 is not necessarily equal to P_0. Consequently, K_1 is not equal to K_0, nor is $P_{1|1}$ equal to $P_{0|0}$. This leads to other questions: do the covariances and the gain converge, after a transient period, to bounded steady-state matrices? If convergence does occur, then the error dynamics of Eq.(A.17) is also an LTI system. Is this system stable?

The answer to these questions is that convergence occurs and the error dynamics are indeed stable whenever the system properties of *reachability* and *observability* from the disturbance are satisfied. These conditions are sufficient but not necessary as will be shown. We first define the system properties just mentioned.

A linear system (A, B), where A is $n \times n$ and B is $n \times m$, is **reachable** if there exists an input history $u_0, ..., u_r, r \leq n - 1$, that would steer the state of the system $x_{k+1} = A x_k + B u_k$ from any initial state x_0, to any other state x. Mathematically, this is equivalent to the condition that the *reachability matrix* $[B, AB, ..., A^{n-1}B]$ be of rank n. Another equivalent condition is that the *reachability Gramian* $\sum_{k=0}^{n-1} A^k BB'(A')^k$ is of rank n.

---

[14]Some authors require the innovation to have a unit covariance. In this case, only the $\check{\varrho}_k$'s would be the innovation of the y_k's.

A linear system (A, C), where A is $n \times n$ and C is $m \times n$, is **observable** if the initial state x_0 can be exactly reconstructed from observations $y_0, ..., y_r, r \leq n - 1$, of the system $x_{k+1} = Ax_k, y_k = Cx_k$. This is equivalent to requiring that the linear map from x_0 to $y_0, ..., y_{n-1}$ be invertible. Note that (A, C) is observable if, and only if, (A', C') is reachable. This leads to two other equivalent conditions: The *observability matrix* $[C', A'C', ..., (A')^{n-1}C')]$ is of rank n, and the *observability Gramian* $\sum_{k=0}^{n-1}(A^k)'C'CA^k$ is of rank n.

Finally, we recall that a linear system is **stable** if the eigenvalues of its dynamic matrix A have magnitude less than unity. We now state the convergence and stability results.

Theorem A.2 *Consider the linear time-invariant system*

$$x_{k+1} = Ax_k + Br_k \tag{A.55}$$
$$y_k = Cx_k + Dr_k \tag{A.56}$$

The vector r_k is uncorrelated in time or white, has zero mean and unit covariance. The initial estimate $\hat{x}_0$ is unbiased with error covariance P_0, and the error in the initial estimate is uncorrelated with the noise process r_k for all k. Assume that (A, B) is reachable, (A, C) is observable, and $CDDC' > 0$. Denote the Kalman filter's one-step prediction error covariance by P_k, then

1. $\lim_{k \to \infty} P_k = P$, *where P is the unique positive definite solution to the discrete-time algebraic Riccati equation*

$$
\begin{aligned}
P &= (A - KC)P(A - KC)' \\
&\quad + (B - KD)(B - KD)'
\end{aligned} \tag{A.57}
$$

with the closed-loop gain K given by

$$K = (APC' + BD')(CPC' + DD')^{-1} \tag{A.58}$$

In particular, P is independent of the initial error covariance P_0.

2. *The steady-state error dynamics $\tilde{x}_{k+1} = (A - KC)\tilde{x}_k + (B - KD)r_k$ are stable, meaning that the eigenvalues of $A - KC$ have magnitude less than unity.*

The convergence of the Riccati equation is discussed in [3], [40], [71], and [88], among others. The one-step predictor is now given by

$$\hat{x}_{k+1} = (A - KC)\hat{x}_k + Ky_k \qquad (A.59)$$

It also follows from the above theorem that the filtered error covariance $P_{k|k}$ converges to a steady-state positive definite matrix P_u given by

$$
\begin{aligned}
P_u &= \lim_{k\to\infty} P_{k|k} \\
&= P - LCP \qquad (A.60)
\end{aligned}
$$

where the open loop gain L is

$$L = PC' \left(CPC' + DD'\right)^{-1} \qquad (A.61)$$

Under the same assumption, all the equations for the gain and error co-variances derived in the previous sections converge as well. The filtered estimate is now given by

$$\hat{x}_{k|k} = (I - LC)\hat{x}_k + Ly_k \qquad (A.62)$$

Note that the above theorem holds even if the plant dynamics matrix A is unstable. Another observation is that reachability and observability are sufficient, but not necessary, conditions for convergence of the Riccati equation, as well as for the stability of the error dynamics. In fact, not even the weaker conditions of stabilizability and detectability are necessary. A system is **stabilizable** if all its unstable modes are reachable, which implies that they can be stabilized. A system is **detectable** if all its unstable modes are observable, or can be reconstructed from the observations.

Consider, for instance, the case where the plant dynamic matrix A is unstable, the process noise is absent so that $B = 0$, the initial state x_0 is known so that $P_0 = 0$, and the observations consist only of noise so that $C = 0$. This system, $x_{k+1} = Ax_k, y_k = Dr_k$, is neither stabilizable nor detectable. Yet, the absence of process noise means that we can propagate the known initial state using the deterministic dynamics in order to obtain the exact value of the state at each step. The resulting covariances are zero!

A second example is the case where A is unstable, the system is not affected by process noise, so that $B = 0$, but $P_0 \neq 0$ and C is chosen so that the pair (A, C) is detectable. This system is not stabilizable. From linear system theory, however, detectability implies that it is possible to find a gain K such that the error dynamic matrix $A - KC$ is stable.

A third example is the case where A is stable, $P_0 \neq 0, B \neq 0$, but the system is not reachable. Moreover, the observations still consist of pure noise, so that $C \equiv 0$ and the system is not observable. In this case, even the suboptimal filter $\hat{x}_k = 0$ produces stable error dynamics, since, as $k \to \infty$, we have $\tilde{x}_k = x_k = 0$!

Finally, we note that for time-varying systems, while the question of convergence does not arise, the issue of stability of the error dynamics does. The sufficient conditions for stability are similar in nature, using definitions of stability, reachability, and observability for time-varying systems.

A.7   The Wiener Filter

The Kalman filter was derived in the late fifties. In this section, we travel back further to the time of Wiener, or the forties. We formulate the Wiener filtering problem, and present its solution. For simplicity, we only discuss the scalar case.

In Chapter 2, the Wiener filtering problem is formulated as follows: Given two wide-sense jointly stationary processes x_k and y_k with known auto- and cross-correlation functions $R_{yy}[k]$ and $R_{xy}[k]$, respectively, the goal is to solve the optimization problem

$$\min_{\hat{x}} \quad E\left(x_k - \hat{x}_k\right)^2 , \quad -\infty < k < +\infty \qquad (A.63)$$

$$\text{with} \quad \hat{x}_k = \sum_{\ell=-\infty}^{k-1} h[k,\ell]y_\ell \qquad (A.64)$$

The optimal filter, which is time-invariant, must satisfy the discrete Wiener Hopf equation [88]

$$R_{xy}[k] = \sum_{\ell=-\infty}^{\infty} h[k-\ell]R_{yy}[\ell] , \quad k \geq 0 \qquad (A.65)$$

The presence of the constraint $k > 0$ in the above equation prevents us from taking the bilateral z or Fourier transform of the equation. A more sophisticated technique is therefore needed, and such a technique was developed by Wiener and Kolmogorov[15]. If $\mathcal{H}_W(z)$, $S_{xx}(z)$, $S_{xy}(z)$, and $S_{yy}(z)$ are the z transform of, respectively, $h[k]$, $R_{xx}[k]$, $R_{xy}[k]$, and $R_{yy}[k]$, then the solution to the Wiener-Hopf equation can be given in terms of these transforms. Specifically,

$$\mathcal{H}_W(z) = \frac{1}{S_{yy}^+(z)} \left\{ \frac{S_{xy}(z)}{S_{yy}^-(z)} \right\}^+ \tag{A.66}$$

The superscript $+$ means that only the causal part of the transform is kept, while the superscript $-$ indicates that only the anticausal portion of the transform is kept[16].

We now discuss how Eq.(A.66) can be derived. As mentioned earlier, the constraint $k > 0$ in Eq.(A.65) prevents us from taking the z or Fourier transform of that equation. If the observation process y is white, however, then Eq.(A.65) would have a simple solution, specifically, $h[k] = R_{xy}[k]$ for $k > 0$, and $h[k] = 0$ for $k \leq 0$. A logical approach, therefore, is to obtain the standard white innovation process ($\breve{\varrho}_k$ in Section A.5) from the observation as a first step, and solve Eq.(A.65) with the innovation replacing the observation as a second step. The first step is simply given by

$$T_{\breve{\varrho}_k y} = \left(S_{yy}^+(z) \right)^{-1} \tag{A.67}$$

For the second step, we have

$$\begin{aligned}
R_{x\breve{\varrho}}[k] &= \sum_{\ell=-\infty}^{\infty} h[k-\ell] R_{\breve{\varrho}_k \breve{\varrho}_k}[\ell] \ , \quad k \geq 0 \\
&= \sum_{\ell=-\infty}^{\infty} h[k-\ell] \delta_\ell \ , \quad k \geq 0
\end{aligned}$$

The solution to the above equation is

$$h[k] = R_{x\breve{\varrho}}[k] u_{-1}[k] \tag{A.68}$$

where $u_{-1}[k]$ is the unit step function. Since h_k must be causal, we can only retain $R_{x\breve{\varrho}}[k]$ for $k > 0$. To do so, we factor the z transform of $R_{x\breve{\varrho}}$ into causal and anti-causal factors, i.e., $S_{x\breve{\varrho}}(z) = S_{x\breve{\varrho}}^+(z) S_{x\breve{\varrho}}^-(z)$, and

---

[15] Some authors refer to the discrete-time Wiener Filter as the Wiener-Kolmogorov filter.

[16] A power spectral density can be factored provided it meets certain technical conditions (Paley-Wiener), which we assume (see [88])

retain only the realizable portion $S_{x\check{\varrho}}^{+}(z)$, which has no poles outside the unit circle. If $H(z)$ is the z-transform of $h[k]$, we then have

$$H(z) = S_{x\check{\varrho}}^{+}(z)$$

As a result,

$$
\begin{aligned}
\mathcal{H}_W(z) &\equiv T_{\hat{x}y}(z) \\
&= T_{\hat{x}\check{\varrho}}(z)T_{\check{\varrho}y}(z) \\
&= H(z)\left(S_{yy}^{+}(z)\right)^{-1} \\
&= S_{x\check{\varrho}}^{+}(z)S_{yy}^{+}(z)
\end{aligned}
\tag{A.69}
$$

We now need express $S_{x\check{\varrho}}(z)$ as a function of $S_{xy}(z)$ and $S_{yy}(z)$. From linear system theory, this expression is given by

$$S_{x\check{\varrho}}(z) = S_{xy}(z)\left(S_{yy}^{-}(z)\right)^{-1} \tag{A.70}$$

Substituting for $S_{x\check{\varrho}}(z)$ from the above equation into Eq.(A.69), we obtain the desired result

$$\mathcal{H}_W(z) = \left\{S_{xy}(z)\left(S_{yy}^{-}(z)\right)^{-1}\right\}^{+}\left(S_{yy}^{+}(z)\right)^{-1} \tag{A.71}$$

That the above expression is equivalent to the Kalman filter can be demonstrated, but requires extensive algebraic manipulations. In [62], this equivalence is demonstrated for a special case of a scalar process.

A.8   Smoothing

In the previous sections of this appendix, we have derived estimates that are functions of past, $\hat{x}_k$, or past and present, $\hat{x}_{k|k}$, observations. In some applications, it may be beneficial to obtain smoothed estimates, or estimates that are functions of past, present and future data [85] and [89]. Smoothed estimates are more accurate than either filtered or predicted estimates, because they are based on more observations.

We shall discuss the following types of smoothing: fixed-interval, fixed-point, and fixed-lag smoothing. *Fixed-interval smoothing* uses a set of observations taken over a finite time interval, and the computations are usually off-line. The state over the entire time interval is to be estimated. This type of smoothing is of interest, for instance, when determining the history of a flight trajectory. *Fixed-point smoothing* uses a finite or infinite time interval, and can be implemented either on-line or off-line. It is concerned with estimates at particular times, such as the initial state of the system. *Fixed-lag smoothing* usually takes place on line, and is concerned with obtaining an improvement over the filtered estimate when a time delay is allowed.

Consider the linear system of Section A.1 with the same assumptions,

$$x_{k+1} = A_k x_k + B_k r_k \qquad (A.72)$$
$$y_k = C_k x_k + D_k r_k \qquad (A.73)$$

Assume that the exogenous disturbance and initial error are Gaussian. The Gaussian assumption is not needed for the formulation of the smoothing problem, but it provides insight. To simplify the derivations, we will assume throughout this section that the process and measurement noise are uncorrelated, or $B_k D_k' = 0, \forall k$. Consider a finite time interval, i.e., $k \in [0, N]$. The fixed-interval smoothing problem is given by

$$\max_{\{x_0, x_1, \dots, x_N\}} p(x_0, x_1, \dots, x_N \mid y_0, \dots, y_N) \qquad (A.74)$$

In contrast, the fixed-point smoothing problem for a finite time interval is formulated as

$$\max_{x_k} p(x_k \mid y_0, \dots, y_N) \quad \forall k \in [0, N] \qquad (A.75)$$

Finally, The fixed-lag smoothing problem, which is usually executed online, is to be formulated over a time interval that is not necessarily finite

$$\max_{x_k} p(x_k \mid y_0, \dots, y_k, y_{k+1}, \dots, y_{k+\ell}) \quad k = 0, 1, 2, \dots \qquad (A.76)$$

where the lag is given by $\ell > 0$. In all three smoothing problems, we are looking for the maximum a posteriori estimate, or MAP. Note that for fixed-interval smoothing as defined by Eq.(A.74), the maximization problem is posed using the joint density function, and its solution provides the maximum a posteriori estimate of the state over an entire interval $\left(\hat{x}_{1|N}, \dots, \hat{x}_{N|N}\right)$. For the fixed-point smoothing problem of Eq.(A.75), however, there are up to N maximization problems, and each uses the *marginal* density. The solution to each of these fixed-point smoothing problems is the maximum a posteriori estimate of the state at a particular time k, which we denote here $\check{x}_{k|N}, k = 0, \dots, N$.

For Gaussian density functions, the fixed-point and fixed-interval smoothing estimates are the same; they are equal to the conditional mean. In general, however, $\hat{x}_{k|N} \neq \check{x}_{k|N}$. This can be illustrated by the following example. Consider two binary random variables, (α_1, α_2), with joint probability mass function

$$p(\alpha_1 = 0, \alpha_2 = 0) = 7/21 \quad , \quad p(\alpha_1 = 1, \alpha_2 = 0) = 3/21$$
$$p(\alpha_1 = 0, \alpha_2 = 1) = 6/21 \quad , \quad p(\alpha_1 = 1, \alpha_2 = 1) = 5/21$$

The marginal probabilities are

$$p(\alpha_1 = 0) = 13/21 \quad , \quad p(\alpha_1 = 1) = 8/21$$
$$p(\alpha_2 = 0) = 10/21 \quad , \quad p(\alpha_2 = 1) = 11/21$$

From the joint and marginal distributions, we see that the fixed-interval and fixed-point smoothed estimates are given respectively by

$$\begin{aligned} \text{From Joint} &\quad : \quad \hat{\alpha}_1 = 0 \quad , \quad \hat{\alpha}_2 = 0; \\ \text{From Marginal} &\quad : \quad \check{\alpha}_1 = 0 \quad , \quad \check{\alpha}_2 = 1 \end{aligned}$$

We now discuss each of the three smoothing problems separately.

A.8.1   Fixed-Interval Smoothing

As mentioned previously, we shall assume that the noise is Gaussian. The Gaussian assumption is not needed, but provides insight into the relationship between the probabilistic and variational approaches to estimation. Define

$$J = \max_{\{x_0, \dots, x_N\}} p(x_0, \dots, x_N \mid y_0, \dots, y_N)$$

The maximization problem of Eq.(A.74) can be rewritten as

$$J = \max_{\{x_0, \dots, x_N\}} \frac{p(y_0, \dots, y_N \mid x_0, \dots, x_N) \, p(x_0, \dots, x_N)}{p(y_1, \dots, y_N)} \tag{A.77}$$

$$= \max_{\{x_0, \dots, x_N\}} p(y_0, \dots, y_N \mid x_0, \dots, x_N) \, p(x_0, \dots, x_N) \tag{A.78}$$

$$= \max_{\{x_0, \dots, x_N\}} \prod_{k=0}^{N} p(y_k | x_k) \prod_{k=1}^{N} p(x_k | x_{k-1}) p(x_0) \tag{A.79}$$

$$= \max_{\{x_0, \dots, x_N\}} \sum_{k=0}^{N} \log p(y_k | x_k) + \sum_{k=1}^{N} \log p(x_k | x_{k-1}) + \log p(x_0)$$

$$(A.80)$$

$$= \min_{\{x_0,\dots,x_N\}} \frac{1}{2} (y_N - C_N x_N)' (D_N D_N')^{-1} (y_N - C_N x_N)$$

$$+ \sum_{k=0}^{N-1} \frac{1}{2} \bigg((y_k - C_k x_k)' (D_k D_k')^{-1} (y_k - C_k x_k)$$

$$+ (x_{k+1} - A_k x_k)' (B_k B_k')^{-1} (x_{k+1} - A_k x_k) \bigg)$$

$$+ \frac{1}{2} (x_0 - \hat{x}_0)' P_0^{-1} (x_0 - \hat{x}_0) \qquad (A.81)$$

The first equality, Eq.(A.77), follows from Bayes' rule. The second equality, Eq.(A.78), follows from the fact that $p(y_1, \dots, y_N)$ does not affect the maximization. The third equality, Eq.(A.79), is obtained by conditioning and by invoking the Markov property of the x_k's, while Eq.(A.80) follows from the fact that the log function is monotone increasing. Finally, the last equality, Eq.(A.81), is a consequence of the Gaussian assumption. For instance,

$$\log p(y_k | x_k) = \log \left(\text{Constant} \times e^{-(y_k - C_k x_k)'(D_k D_k')^{-1}(y_k - C_k x_k)} \right)$$

$$= \text{Constant} - (y_k - C_k x_k)' (D_k D_k)^{-1} (y_k - C_k x_k)$$

and

$$\log p(x_{k+1} | x_k) = \text{Constant} - (x_{k+1} - A_k x_k)' (B_k B_k')^{-1} (x_{k+1} - A_k x_k)$$

Note that the maximization turned into a minimization in Eq.(A.81) because the exponent of the Gaussian is negative.

The minimization of Eq.(A.81) can be reformulated as a variational, or a constrained, optimization problem in the decision variables x_0 and r_k, $k = 0, 1, \dots, N$. Specifically, we have

$$\min_{\{x_0, r_0, \dots, r_N\}} J \qquad (A.82)$$

where

$$J \equiv \frac{1}{2} (x_0 - \hat{x}_0)' P_0^{-1} (x_0 - \hat{x}_0)$$

$$+ \frac{1}{2} \sum_{k=0}^{N-1} \bigg((y_k - C x_k)' (D_k D_k')^{-1} (y_k - C x_k) + r_k' r_k \bigg)$$

$$+ \frac{1}{2} (y_N - C_N x_N)' (D_N D_N')^{-1} (y_N - C_N x_N) \qquad (A.83)$$

subject to the constraints

$$x_{k+1} = A_k x_k + B_k r_k \quad \forall k \in [0, N-1] \tag{A.84}$$

The Hamiltonian matrix equation for the above variational problem is

$$\begin{bmatrix} x_{k+1}^s \\ -\lambda_k^s \end{bmatrix} = \begin{bmatrix} A_k & B_k B_k' \\ C_k' (D_k D_k')^{-1} C_k & -A_k' \end{bmatrix} \begin{bmatrix} x_k^s \\ \lambda_{k+1}^s \end{bmatrix}$$
$$+ \begin{bmatrix} 0 \\ -C_k' (D_k D_k')^{-1} \end{bmatrix} y_k \tag{A.85}$$

with boundary conditions

$$x_0^s = \hat{x}_0 + P_0 \lambda_0 \tag{A.86}$$
$$\lambda_{N+1} \quad \text{free} \tag{A.87}$$

The superscript s indicates that the estimate is smoothed. To obtain a recursive formulation for the smoother, the Hamiltonian matrix can be triangularized or diagonalized using the transformation

$$\hat{x}_k = x_k^s - P_k \lambda_k^s \tag{A.88}$$

where $\hat{x}_k$ is the one-step prediction estimate and P_k is the corresponding prediction error covariance derived in Section A.2. After some matrix algebra, this gives

$$\begin{bmatrix} \hat{x}_{k+1} \\ -\lambda_k^s \end{bmatrix} = \begin{bmatrix} A_k - K_k C_k & 0 \\ \check{F}_k C_k' (D_k D_k')^{-1} C_k & -\check{F}_k A_k' \end{bmatrix} \begin{bmatrix} \hat{x}_k \\ \lambda_{k+1}^s \end{bmatrix}$$
$$+ \begin{bmatrix} K_k \\ -\check{F}_k C_k' (D_k D_k')^{-1} \end{bmatrix} y_k \tag{A.89}$$
$$= \begin{bmatrix} A_k & 0 \\ 0 & -\check{F}_k A_k' \end{bmatrix} \begin{bmatrix} \hat{x}_k \\ \lambda_{k+1}^s \end{bmatrix}$$
$$+ \begin{bmatrix} K_k \\ -\check{F}_k C_k' (D_k D_k')^{-1} \end{bmatrix} \varrho_k \tag{A.90}$$

with boundary conditions

$$\hat{x}_0 \quad \text{given} \tag{A.91}$$
$$\lambda_{N+1}^s \quad \text{free} \tag{A.92}$$

where ϱ_k is the innovation, and

$$
\begin{aligned}
K_k &= A_k \left(P_k^{-1} + C_k' \left(D_k D_k' \right)^{-1} C_k \right)^{-1} C_k' \\
&= A_k P_k C_k' \left(C_k P_k C_k' + D_k D_k' \right)^{-1} \quad\quad\quad\text{(A.93)} \\
P_{k+1} &= \left(A_k - K_k C_k \right) P_k \left(A_k - K_k C_k \right)' \\
&\quad + \left(B_k - K_k D_k \right) \left(B_k - K_k D_k \right)' \quad\quad\text{(A.94)} \\
\check{F}_k &= \left(-I + C_k' \left(D_k D_k' \right)^{-1} C_k P_k \right)^{-1} \quad\quad\text{(A.95)}
\end{aligned}
$$

We note again that the top row of this new Hamiltonian equation, together with the equations for K_k, P_k, and the given initial value for $\hat{x}_0$, form the Kalman filter. After more matrix algebra, it is possible to obtain the following alternate diagonal form for the Hamiltonian equation

$$
\begin{bmatrix} \hat{x}_{k+1} \\ -\lambda_k^s \end{bmatrix} = \begin{bmatrix} A_k & 0 \\ 0 & -\left(A_k - K_k C_k \right)' \end{bmatrix} \begin{bmatrix} \hat{x}_k \\ \lambda_{k+1}^s \end{bmatrix}
$$
$$
+ \begin{bmatrix} K_k \\ -C_k' \left(C_k P_k C_k' + D_k D_k' \right)^{-1} \end{bmatrix} \varrho_k \quad\quad\text{(A.96)}
$$

with the same boundary conditions. The smoother is obtained by first obtaining a one-step prediction estimate from the forward sweep. Once $\hat{x}_N$ is obtained, a backward sweep using the bottom row of any of the above hamiltonian equation in $\hat{x}_k$ gives the values of λ_k^s. If we assume that for the fictitious step from N to $N+1$, we have $A_N = I$ and $B_N = 0$, then $\hat{x}_{N|N} = \hat{x}_{N+1}$. We also have $\hat{x}_{N|N} = \hat{x}_N^s$. With the last measurement taken at time N, it then follows that $\hat{x}_{N+1}^s = \hat{x}_N^s = \hat{x}_{N+1}$. As a result, the backward sweep can be started with the guess $\lambda_{N+1}^s = 0$. The smoothed estimate x^s follows from Eq.(A.88). It is also possible to derive the error covariance P_k^s for the fixed-interval smoothed estimate and obtain

$$
P_k^s = P_k - P_k Z_k P_k \quad\quad\text{(A.97)}
$$

where Z_k obeys the recursion (with $Z_{N+1} = 0$)

$$
Z_k = \left(A_k - K_k C_k \right)' Z_{k+1} \left(A_k - K_k C_k \right) + C_k' \left(C_k P_k C_k' + D_k D_k' \right)^{-1} C_k
$$
$$
\text{(A.98)}
$$

Note that, due to the Gaussian noise assumption, the solution to the above variational problem provides the mean and error covariance of the state, and therefore the entire conditional density function. We shall not show it, but if the noise is not Gaussian, then the solution to the above variational problem provides the *linear* minimum variance fixed-interval smoother.

A.8.2   Fixed-Point and Fixed-Lag Smoothing

If the noise is Gaussian, then the solution to the fixed-interval smoothing problem gives the entire joint conditional density function $p(x_0, ..., x_N \mid y_0, ..., y_N)$. We can then obtain the marginal distributions $p(x_j \mid y_0, ..., y_N)$ for any desired $j \in [0, N]$. The mean and covariance of that density function give the solution to the fixed-point smoothing estimate for a finite interval.

There is also a direct approach to the fixed-point problem that can be applicable to the non-Gaussian case. For a desired $j \in [0, N]$, define, for $j \leq k$, the augmented plant and observations,

$$\begin{bmatrix} x_{k+1} \\ x_{j,k+1} \end{bmatrix} = \begin{bmatrix} A_k & 0 \\ 0 & I \end{bmatrix} \begin{bmatrix} x_k \\ x_{j,k} \end{bmatrix} + \begin{bmatrix} B_k \\ 0 \end{bmatrix} r_k \qquad (A.99)$$

$$y_k = \begin{bmatrix} C_k & 0 \end{bmatrix} \begin{bmatrix} x_k \\ x_{j,k} \end{bmatrix} \qquad (A.100)$$

with estimates and error covariance at time $k = j$ given by

$$\begin{bmatrix} \hat{x}_j \\ \hat{x}_j \end{bmatrix} \quad , \quad \begin{bmatrix} P_j & P_j \\ P_j & P_j \end{bmatrix} \qquad (A.101)$$

Applying the Kalman filter to the system of Eqs.(A.99-A.101) gives $\hat{x}_{j,k}$ and $\hat{x}_{j,k|k}$, the fixed-point smoothed estimate of the state at time j as a function of the observations up to, respectively, time $k - 1$ and k, as well as the filtered estimate for $k \geq j$, or $\hat{x}_k$.

For fixed-lag smoothing, we use the augmented system

$$\begin{bmatrix} x_{k+1} \\ x_{1,k+1} \\ x_{2,k+1} \\ \vdots \\ x_{\ell+1,k+1} \end{bmatrix} = \begin{bmatrix} A_k & 0 & \cdots & 0 \\ I & 0 & \cdots & 0 \\ 0 & I & \cdots & 0 \\ \vdots & \vdots & \ddots & \vdots \\ 0 & 0 & \cdots & I \end{bmatrix} \begin{bmatrix} x_k \\ x_{1,k} \\ x_{2,k} \\ \vdots \\ x_{\ell+1,k} \end{bmatrix} + \begin{bmatrix} B_k \\ 0 \\ 0 \\ \vdots \\ 0 \end{bmatrix} r_k$$

$$(A.102)$$

$$y_k = \begin{bmatrix} C_k & 0 & \cdots & 0 \end{bmatrix} \begin{bmatrix} x_k \\ x_{1,k} \\ x_{2,k} \\ \vdots \\ x_{\ell+1,k} \end{bmatrix} + D_k r_k \quad \text{(A.103)}$$

and apply the Kalman filter. The one-step prediction estimate of the state of the above augmented system contains the fixed-lag estimate. It is clear that $\hat{x}_{i+1,k+1} = \hat{x}_{k-i|k}$, for $i = 0, \ldots, \ell$. The associated covariance matrix can be augmented as is done for the fixed-point problem.

For both the fixed-point and the fixed-lag smoothed estimate, most of the knowledge about the state of the system comes from the filter. This is due to the fact that the dynamic system $x_{k+1} = A_k x_k + B_k r_k$ is causal, and therefore, the information about the state is burried mostly in the past. Nevertheless, it is possible to improve the state estimate by incorporating future observations for up to two or three time constants.

A.9 The Extended Kalman Filter (EKF)

In Sections A.6 and A.7, we focused on linear time-invariant systems at steady-state. These systems are special cases of the linear time-varying systems for which we derived recursive estimators. In this section, we make a move in the opposite direction. We consider the more general class of nonlinear systems, or systems whose dynamics and observations are nonlinear functions of the state. The optimal filter equations for nonlinear systems, which we derive, are difficult to compute. As an approximation, the extended Kalman filter (EKF) is used. This algorithm linearizes the nonlinear dynamic and observations around the current estimate at each time step.

Consider the nonlinear plant and observation equations

$$x_{k+1} = f(x_k) + b(x_k)u_k + g(x_k)r_k \quad \text{(A.104)}$$
$$y_k = h(x_k) + D_k r_k \quad \text{(A.105)}$$

where r_k is a white, zero-mean, unit variance disturbance signal. The system of Eq.(A.104) is Markov, so that $p(x_{k+1}|x_k, x_{k-1}, \ldots) = p(x_{k+1}|x_k)$. We will now derive the prediction and update form for the entire density function. The one step prediction is given by

$$p(x_{k+1}|y_0, \ldots, y_k) = \int p(x_{k+1}|x_k, y_0, \ldots, y_k) p(x_k|y_0, \ldots, y_k) \, dx_k$$

$$= \int p(x_{k+1}|x_k) \, p(x_k|y_0, ..., y_k) \, dx_k \qquad \text{(A.106)}$$

The first equality is obtained by conditioning on x_k, and then averaging over all its possible values. The second equality says that, given the actual value of the state x_k, then x_{k+1} is independent of the past observations since the x_k's are Markov. Thus, we used conditioning and the Markov property to obtain propagation equation for the entire density function. For the update equation, we use Bayes' rule

$$p(x_k|y_0, y_1, ..., y_k) = \frac{p(y_k|x_k, y_0, ..., y_{k-1}) \, p(x_k|y_0, ..., y_{k-1})}{\int p(y_k|x_k, y_0, ..., y_{k-1}) \, p(x_k|y_0, ..., y_{k-1}) \, dx_k}$$

$$= \frac{p(y_k \mid x_k) \, p(x_k|y_0, ..., y_{k-1})}{p(y_k|y_0, ..., y_{k-1})} \qquad \text{(A.107)}$$

One can combine the prediction and update equations by substituting Eq.(A.107) into Eq.(A.106), or vice versa. Unfortunately, the density functions in these equations are not easy to compute, a fact that limits their usefulness. The most common approach used to circumvent this difficulty is through the linearization of the plant dynamics and measurements around the current estimate $\hat{x}_k$ or $\hat{x}_{k|k}$. The Kalman filter can then be applied to the linearized system. Specifically, we have the Taylor expansion

$$f(x_k) = f\left(\hat{x}_{k|k}\right) + \frac{\partial}{\partial x_k} f\left(\hat{x}_{k|k}\right) \left(x_k - \hat{x}_{k|k}\right) + \ldots$$

$$= f\left(\hat{x}_{k|k}\right) + A_k \left(x_k - \hat{x}_{k|k}\right) + \ldots$$

$$g(x_k) = g\left(\hat{x}_{k|k}\right) + \frac{\partial}{\partial x_k} g\left(\hat{x}_{k|k}\right) \left(x_k - \hat{x}_{k|k}\right) + \ldots$$

$$= G_k + \ldots$$

$$b(x_k) = b\left(\hat{x}_{k|k}\right) + \frac{\partial}{\partial x_k} b\left(\hat{x}_{k|k}\right) \left(x_k - \hat{x}_{k|k}\right) + \ldots$$

$$= B_k + \ldots$$

$$h(x_k) = h(\hat{x}_k) + \frac{\partial}{\partial x_k} h(\hat{x}_k) (x_k - \hat{x}_k) + \ldots$$

$$= h(\hat{x}_k) + C_k + \ldots$$

Substituting from the above equations into Eqs.(A.104-A.105), and neglecting higher order terms, we have

$$x_{k+1} = A_k x_k + B_k u_k + G_k r_k + \zeta_k \qquad \text{(A.108)}$$

$$y_k = C_k x_k + D_k r_k + \varepsilon_k \qquad \text{(A.109)}$$

where ζ_k and ε_k are known inputs given by

$$\zeta_k = f(\hat{x}_{k|k}) - A_k \hat{x}_{k|k} \tag{A.110}$$

$$\varepsilon_k = h(\hat{x}_k) - C_k \hat{x}_k \tag{A.111}$$

For simplicity, we shall assume that $B_k D'_k = 0$ for all k. A variation of the Kalman filter can then be applied to the systems of Eqs.(A.108-A.109). The resulting filter is called the *extended Kalman filter*, or EKF, with equations given by

$$\hat{x}_k = f(\hat{x}_{k|k}) + B_k u_k \tag{A.112}$$

$$\hat{x}_{k|k} = \hat{x}_k + L_k (y_k - h(\hat{x}_k)) \tag{A.113}$$

with initial conditions $\hat{x}_0$. The remaining equations are exactly those of the Kalman filter. Specifically,

$$L_k = P_k C'_k (C_k P_k C'_k + D_k D'_k)^{-1} \tag{A.114}$$

$$P_{k|k} = P_k - P_k C'_k (C_k P_k C'_k + D_k D'_k)^{-1} C_k P_k \tag{A.115}$$

$$P_{k+1} = (A_k - A_k L_k C_k) P_k (A_k - A_k L_k C_k)' \\ + (G_k - A_k L_k D_k)(G_k - A_k L_k D_k)' \tag{A.116}$$

It is important to note that even though linearization permits the use of the Kalman filter equations, this extension comes at a price, as *none* of the properties of the Kalman filter estimates are guaranteed. Specifically, since the linearization introduces errors, we no longer have a minimum variance estimate, and the matrices P_k and $P_{k|k}$ are no longer the error covariances, but approximations thereof. They are simply Riccati matrices. The quality of the estimates depends in part on the size of the neglected terms in the Taylor series expansions. It is also possible to keep higher order terms in the expansions, and design filters based on the resulting approximations of the plant model. A discussion on the properties of the EKF can be found in [88], where higher order filters are also derived.

When the system is linear, it is possible to compute the covariance matrices off-line and store them. This is not possible here since we have to wait for the estimate at each step in order to determine what the linearized plant model is. However, this disadvantage is not serious due to the fact that computers will only become more and more powerful. If off-line computation is necessary, then it is possible to precompute the matrices P_k and $P_{k|k}$ by linearizing the plant and observation models around a nominal trajectory.

A.10   Summary of Equations and Additional Remarks

1. **The Kalman filter equations**: For the linear system of Eqs.(A.1-A.2) and the accompanying assumptions, we have the following equations for the filter:

   - One-step prediction:

$$\hat{x}_{k+1} = (A_k - K_k C_k)\,\hat{x}_k + K_k y_k \tag{A.117}$$

$$K_k = (A_k P_k C_k' + B_k D_k')\,(C_k P_k C_k' + D_k D_k')^{-1} \tag{A.118}$$

$$\begin{aligned} P_{k+1} = &\ (A_k - K_k C_k)\,P_k\,(A_k - K_k C_k)' \\ &+ (B_k - K_k D_k)\,(B_k - K_k D_k)' \end{aligned} \tag{A.119}$$

   with initial conditions $\hat{x}_0$ and P_0. If the measurement and process noise are uncorrelated, so that $B_k D_k' \equiv 0$ for all k, then it is possible to express the one-step predicted estimate and error covariance in terms of the filtered estimate and error covariance. Specifically,

$$\hat{x}_{k+1} = A_k \hat{x}_{k|k} \tag{A.120}$$

$$P_{k+1} = A_k P_k A_k' + B_k B_k' \tag{A.121}$$

   - Measurement update:

$$\hat{x}_{k|k} = (I - L_k C_k)\,\hat{x}_k + L_k y_k \tag{A.122}$$

$$L_k \equiv P_k C_k' \,(C_k P_k C_k' + D_k D_k')^{-1} \tag{A.123}$$

$$P_{k|k} = P_k - P_k C_k' \,(C_k P_k C_k' + D_k D_k')^{-1} C_k P_k \tag{A.124}$$

2. **Colored Noise:** If the input disturbance includes colored noise, then the framework we used is still applicable, provided the power spectral density of the noise is known, and satisfies some technical conditions (See [88]), which we assume. Specifically, let ζ_k be a colored input disturbance process with spectral density $S_{\zeta\zeta}$. The plant and sensor equations are now

$$x_{k+1} = A_k x_k + B_k r_k + B_\zeta \zeta_k \tag{A.125}$$

$$y_k = C_k x_k + D_k r_k + D_\zeta \zeta_k \tag{A.126}$$

Assume it is possible to obtain the factorization $S_{\zeta\zeta} = H_\zeta(z)H_\zeta(z^{-1})$, where $H_\zeta(z)$ is a rational transfer function. Let the linear system realization of $H_\zeta(z)$ be

$$\begin{aligned}
\check{r}_{k+1} &= A_\zeta \check{r}_k + \Gamma_\zeta \varepsilon_k \\
\zeta_k &= C_\zeta \check{r}_k
\end{aligned}$$

where $\check{r}_k$ is a white noise process independent of, and with the same statistical properties as r_k. Then we can recover the framework we used in this appendix with the following substitutions in Eqs.(A.1-A.2)

$$x_k \leftarrow \begin{bmatrix} x_k \\ \check{r}_k \end{bmatrix} \quad , \quad r_k \leftarrow \begin{bmatrix} r_k \\ \varepsilon_k \end{bmatrix}$$

$$A_k \leftarrow \begin{bmatrix} A_k & B_\zeta C_\zeta \\ 0 & A_\zeta \end{bmatrix} \quad , \quad B_k \leftarrow \begin{bmatrix} B_k & 0 \\ 0 & \Gamma_\zeta \end{bmatrix}$$

$$C_k \leftarrow \begin{bmatrix} C_k & D_\zeta C_\zeta \end{bmatrix} \quad , \quad D_k \leftarrow D_k$$

3. **Least Squares Error (LSE) Estimation** and **Minimum Mean-Squares Error Estimation**: These terms are often used in the literature to mean minimum variance estimation. They also imply that the Kalman filter minimizes any weighted Euclidean norm of the error vector, or $\|\tilde{x}\|_Q$, where Q is any positive, semi-definite weight matrix. This deterministic property of the Kalman filter is seen in Chapter 3.

4. **Recursive Least Squares (RLS) Algorithm**: Section A.7 took us back in time from the days of Kalman to the days of Wiener. The following remark takes us back to the days of Gauss. A special case of the Kalman filter is the well-known RLS algorithm. This algorithm is used to recursively estimate a static quantity using noisy observations. In our context, this problem is formulated with the linear model of Eqs.(A.1-A.2) by setting $A_k \equiv I$ and $B_k \equiv 0$, for all k. Specifically, we have

$$\begin{aligned}
x_{k+1} &= x_k \\
y_k &= C_k x_k + D_k r_k
\end{aligned}$$

The equations for the RLS algorithm can then be obtained from the Kalman filter equation. Notice that, since the plant is static, there is only an update equation. The RLS algorithm is then just the update equations of the Kalman filter with the substitutions $\hat{x}_k \leftarrow \hat{x}_{k-1|k-1}$ and $P_k \leftarrow P_{k-1|k-1}$. Thus, we have

$$\hat{x}_{k|k} = (I - L_k C_k)\hat{x}_{k-1|k-1} + L_k y_k \qquad (A.127)$$

$$L_k = P_{k-1|k-1}C_k' \left(C_k P_{k-1|k-1}C_k' + D_k D_k'\right)^{-1} \qquad (A.128)$$

$$= P_{k|k}C_k'(D_k D_k')^{-1} \qquad (A.129)$$

$$P_{k|k} = P_{k-1|k-1} -$$

$$P_{k-1|k-1}C_k' \left(C_k P_{k-1|k-1}C_k' + D_k D_k'\right)^{-1} C_k P_{k-1|k-1}$$

$$\qquad (A.130)$$

$$= (I - L_k C_k) P_{k-1|k-1} \qquad (A.131)$$

The RLS algorithm is used extensively in system identification in order to estimate a plant's unknown parameters ([6],[77]).

5. Even for the case of non-Gaussian disturbance, we did not need to assume the exact form of Eq.(A.12). The only assumption needed is that the estimate be a linear function of the observations available. The form then follows. Assuming the form, however, makes the derivation easier. In any case, as we said before, when the disturbance is Gaussian, no assumption on the form of the estimator is needed.

OUTPUTS OF LINEAR SYSTEMS AND THEIR QUADRATIC FORMS

This appendix describes methods for computing the probability density functions of outputs of linear systems and their quadratic forms. When the system inputs are Gaussian, so are the outputs. In Section B.1, expressions for the mean and covariance of outputs of linear systems in terms of the system parameters are obtained, while in Section B.2, a method is shown for computing the p.d.f.'s of quadratic forms of these Gaussian outputs. These p.d.f.'s were used in Chapters 4 and 5, where their plots are shown.

B.1   Moments of Linear Systems Outputs

Expressions for the moments are derived for linear time-invariant systems, but the method is equally applicable to time-varying systems. Consider the system

$$
\begin{aligned}
x_{k+1} &= A x_k + B d_k \\
y_k &= C x_k + D d_k
\end{aligned}
$$

with estimator equation

$$
\begin{aligned}
\hat{x}_k &= A_e \hat{x}_k + K y_k \\
&= A_e \hat{x}_k + K C x_k + K D d_k
\end{aligned}
$$

We assume that the inputs d_k are Gaussian with mean α_k and unit covariance. The goal is to derive the density function of a sequence of outputs

$$z = [z_0, ..., z_{K-1}]$$

where

$$z_k = M(x_k - \hat{x}_k)$$

It is clear that z is Gaussian. To determine its mean and covariance, consider the augmented system:

$$
\begin{aligned}
x_{a\,k+1} &= \begin{bmatrix} x_{k+1} \\ \hat{x}_{k+1} \end{bmatrix} \\
&= \begin{bmatrix} A & 0 \\ KC & A_e \end{bmatrix} \begin{bmatrix} x_k \\ \hat{x}_k \end{bmatrix} + \begin{bmatrix} B \\ KD \end{bmatrix} d_k \\
&= A_a x_{ak} + B_a d_k
\end{aligned}
$$

Assume that x_{a0} has mean β_0 and covariance P_0. Then the first moments β_k of x_{ak} are determined recursively from:

$$\beta_{k+1} = A_a \beta_k + B_a \alpha_k$$

If $M_a = \begin{bmatrix} M & -M \end{bmatrix}$, then the first moment of z is:

$$
\mu = \begin{bmatrix} \mu_1 \\ ... \\ \mu_K \end{bmatrix}
$$

where

$$\mu_k = M_a x_{ak} \tag{B.1}$$

To obtain the covariance matrix of x_{ak}, use the Lyapunov equation:

$$P_k = A_a P_{k-1} A_a' + B_a B_a' \tag{B.2}$$

To obtain the time correlations, note that

$$
\begin{aligned}
P_{k+1,k} &\equiv E\left((x_{a,k+1} - \beta_{k+1})(x_{ak} - \beta_k)'\right) \\
&= E\left((Ax_{ak} + Bd_k - A\beta_k - B\alpha_k)(x_{ak} - \beta_k)'\right) \\
&= E\left(A(x_{ak} - \beta_k)(x_{ak} - \beta_k)'\right) + E\left((Bd_k - B\alpha_k)(x_{ak} - \beta_k)'\right)
\end{aligned}
$$

The first term in the last equality is simply $A_a P_k$, while the second term is 0 since d is a white process, so that d_k is independent of x_{ak}. Therefore

$$P_{k+1,k} = A_a P_k$$

More generally,

$$\begin{aligned} P_{k+j,k} &\equiv E\left(x_{k+j} x_k'\right) \\ &= A_a^j P_k \end{aligned}$$

and

$$\begin{aligned} P_{k,k+j} &\equiv P_{k+j,k}' \\ &= P_k A_a'^j \end{aligned}$$

The covariance and correlation matrices for each z_k are then:

$$\begin{aligned} Q_k &= M_a P_k M_a' \\ Q_{k+j,k} &= M_a P_{k+j,k} M_a' \\ &= M_a A_a^j P_k M_a' \\ Q_{k,k+j} &= M_a P_k A_a'^j M_a' \end{aligned}$$

All this gives the covariance for the entire vector z:

$$\begin{aligned} \mathcal{Q} &= \begin{bmatrix} Q_1 & \cdots & Q_{1,k} & \cdots & Q_{1,K} \\ & & \cdots & & \\ Q_{k,1} & \cdots & Q_{k,k} & \cdots & Q_{k,K} \\ & & \cdots & & \\ Q_{K,1} & \cdots & Q_{K,k} & \cdots & Q_K \end{bmatrix} \\[2mm] &= \begin{bmatrix} M_a P_1 M_a' & \cdots & M_a P_1 A'^{k-1} M_a' & \cdots & M_a P_1 A'^{K-1} M_a' \\ & & \cdots & & \\ M_a A^{k-1} P_1 M_a' & \cdots & M_a P_k M_a' & \cdots & M_a P_k A'^{K-k} M_a' \\ & & \cdots & & \\ M_a A^{K-1} P_1 M_a' & \cdots & M_a A^{K-k} P_k M_a' & \cdots & M_a P_K M_a' \end{bmatrix} \end{aligned}$$

$$\text{(B.3)}$$

B.2   Probability Density Functions of Gaussian Quadratic Forms

The goal is to derive the probability density function of the quadratic form

$$v = z' \Sigma z \tag{B.4}$$

when z is a Gaussian random vector with mean μ and covariance Q. To do so, consider the singular value decomposition of the covariance matrix Q

$$Q = L'\Lambda L$$

where L is orthogonal and Λ is the matrix of singular values of Q, and define

$$w = \left(L'\Lambda^{\frac{1}{2}}\right)^{-1} z$$

Then, w is Gaussian with mean $\nu = \left(L'\Lambda^{\frac{1}{2}}\right)^{-1} \mu$ and unit covariance. Moreover

$$
\begin{aligned}
v &= z'\Sigma z \\
&= w'Sw \\
&= w'L'_s RL_s w
\end{aligned}
$$

where

$$S = \Lambda^{\frac{1}{2}} L'\Sigma L\Lambda^{\frac{1}{2}}$$

and $L'_s RL_s$ is the singular value decomposition of S, with $R = diag(\rho_1, ..., \rho_K)$. Since L_s is unitary with respect to the matrix 2-norm, then v has the same density function as $w'Rw$ That is,

$$
\begin{aligned}
v &\overset{d}{=} w'Rw \\
&= \sum_{k=1}^{K} \rho_k w_k^2 \\
&= \sum_{k=1}^{K} \rho_k t_k
\end{aligned}
$$

where $\overset{d}{=}$ stands for equal in distribution, and the t_k's are independent random variable,each having a standard noncentral χ^2 density function with one degree of freedom and noncentrality parameter ν_k^2, the kth component of ν squared:

$$f_{t_k}(t) = \frac{1}{\sqrt{2\pi t}} e^{-(t+\nu_k^2)/2} \cosh\left(\sqrt{t}\nu_k\right)$$

If $\mathcal{F}$ is the Fourier transform operation, then the corresponding characteristic function is:

$$
\begin{aligned}
\Psi_{t_k}(\omega) &= \mathcal{F}(f_{t_k}) \\
&= \frac{1}{(1-2i\omega)^{\frac{1}{2}}} \exp \frac{i\omega\nu_k^2}{1-2i\omega}
\end{aligned}
$$

Thus, the density of v is that of a weighted sum of independent random variables, or equivalently

$$
\begin{aligned}
f_v(t) &= \frac{1}{\rho_1} f_{t_1}(t/\rho_1) * \ldots * \frac{1}{\rho_K} f_{t_K}(t/\rho_K) \\
&= \mathcal{F}^{-1}\left\{ \prod_{k=1}^{K} \frac{1}{\rho_k} \mathcal{F}\left[f_{t_k}(t/\rho_k)\right] \right\} \tag{B.5} \\
&= \mathcal{F}^{-1}\left\{ \prod_{k=1}^{K} \left[\frac{1}{(1 - 2\rho_k i\omega)^{\frac{1}{2}}} \exp\left(\frac{i\omega \nu_k^2 \rho_k}{1 - 2\rho_k i\omega} \right) \right] \right\} \tag{B.6}
\end{aligned}
$$

Expression B.5 or B.6 can be used to compute the pdf's of quadratic forms in Chapters 4 and 5 with the help of numerical FFT and IFFT algorithms.

APPENDIX C
CONTINUOUS-TIME ROBUST ESTIMATION

C.1 Introduction

In this appendix, estimators are derived for continuous-time linear systems that are robust to a general class of model uncertainties. These estimators are the continuous-time analog of the discrete ones derived in Chapter 3. They are applicable to time-varying and time-invariant systems defined over a finite or infinite-time horizon, with an arbitrary initial condition. Since the approach used here parallels that of the discrete case, we will address the robust estimation problem directly, i.e., the continuous analog of the problem of Section 3.3. As is the case with discrete-time systems, the steady-state estimator equations have the same form as the finite-time horizon estimator, and will not be given.

The results in this chapter can be extended to linear time-invariant systems at steady state, as is shown in Section 3.4 for the discrete case. Moreover, smoother equations similar to those of Section 3.5 can also be derived.

In the next section, the robust estimation problem is formulated. The derivation of the estimator equations is presented in Section C.3, where the estimator properties are also discussed. The development described in Chapter 3 is repeated here so that the chapter may be independently read, but proofs of theorems are not given, as the ideas behind them are similar to those of the discrete estimation problem.

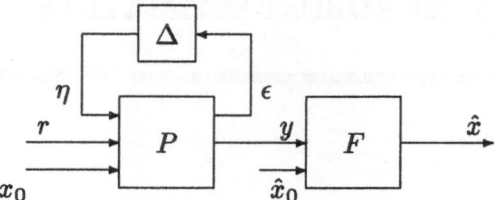

Figure C.1: General representation of robust estimation problem

C.2   Problem Formulation

Figure C.1 shows a general input/output representation of a nominal plant P with modeling uncertainties Δ and an estimator F. All inputs and outputs are real vectors. For $t \in [0, t_f]$, the exogenous input to the plant is the vector $r(t)$, which includes both process and measurement noise, the actual initial state x_0 of the plant, and the estimated initial state $\hat{x}_0$. Both the initial condition and the initial error $x_0 - \hat{x}_0$ are assumed to lie in a set defined by a Euclidean norm bound. The plant output is the measurement vector $y(t)$. This output is in turn fed to the estimator F, whose output is the estimation error, $e(t) \equiv M(t)(x(t) - \hat{x}(t))$. Finally, $\epsilon(t)$ and $\eta(t)$ represent the signals connecting the nominal plant and the perturbation. The signals $\eta(t)$, $r(t)$, $y(t)$, and $\epsilon(t)$ are assumed to lie in the space $L_2[0, t_f]$ of square integrable functions. The norm of a signal $s \in L_2[0, t_f]$ is defined as

$$\|s\| \equiv \left(\int_0^{t_f} s'(t)s(t)dt \right)^{1/2}$$

In addition, we denote the weighted Euclidean norm by $\|.\|_Q$, where Q is a symmetric positive definite matrix of appropriate dimension.

The robust estimator seeks to bound the induced norm of the operator from the input disturbances r and initial estimation error $(x_0 - \hat{x}_0)$ to the estimation error e, for all possible model perturbations Δ of bounded induced 2-norm. Let $\mathcal{G}$ be the mapping between the input disturbance r and initial state estimation error $x_0 - \hat{x}_0$ to the estimation error e. Mathematically, we would like to satisfy the following performance criterion:

$$\|\bar{\mathcal{G}}\|_{i2}^2 \equiv \sup_{\{\|r\|^2+\|x_0-\hat{x}_0\|_{P_0^{-1}}^2+\|x_0\|_{\bar{X}_0}^2=1\}} \|e\|^2 < 1$$

$$\forall \Delta \ni \quad \|\Delta\|_{i2}^2 \equiv \sup_{\|\epsilon\|\neq 0} \frac{\|\eta\|^2}{\|\epsilon\|^2} < 1 \tag{C.1}$$

where the bounds are set to one assuming rescaling is possible. The weighting matrix P_0, which is assumed to be symmetric and positive definite, is a measure of the uncertainty in the initial guess.

The approach used to achieve the performance goal of Eq.(C.1) consists of treating the perturbation output η as an additional exogenous disturbance input and the perturbation input ϵ as an additional plant output. We then find an estimator whose objective is to bound the induced norm of the mapping from the augmented input to the augmented output by one. This will guarantee that the performance objective of Eq.(C.1) is achieved. A new performance criterion is defined as

$$\sup_{(r,\eta,x_0,x_0-\hat{x}_0)\neq 0} J_1 < 1 \tag{C.2}$$

where

$$J_1 \equiv \frac{\|e\|^2 + \|\epsilon\|^2}{\|r\|^2 + \|\eta\|^2 + \|x_0\|_{\bar{X}_0}^2 + \|x_0 - \hat{x}_0\|_{P_0^{-1}}^2}$$

Eq.(C.2) is a robust performance or small gain condition. Proposition 3.3, rewritten below, states that if this criterion is satisfied, then the original performance condition of Eq.(C.1) is also satisfied for the entire class of bounded perturbations Δ.

Proposition C.1 *If 1)* $\bar{J}_1 < 1$, *and 2)* $\|\Delta\|^2 < 1$, *then* $\|\bar{\mathcal{G}}\|_{i2}^2 < 1$.

With $d(t) = [r(t)'\ \eta(t)']'$, we assume the nominal plant and estimator have the linear time-varying state-space representation

$$\begin{aligned}
\dot{x}(t) &= A(t)x(t) + B(t)d(t) \\
\epsilon(t) &= S(t)x(t) + T(t)d(t) \\
e(t) &= M(t)(x(t) - \hat{x}(t)) \\
y(t) &= C(t)x(t) + D(t)d(t)
\end{aligned}$$

$$\begin{bmatrix} \dot{x} \\ \epsilon \\ e \\ y \end{bmatrix}(t) = \left[\begin{array}{c|cc} A & B & 0 \\ S & T & 0 \\ \hline M & 0 & -M \\ C & D & 0 \end{array}\right](t) \begin{bmatrix} x \\ \hline d \\ \hat{x} \end{bmatrix}(t) \tag{C.3}$$

with initial condition x_0 and initial error $e_0 \equiv x_0 - \hat{x}_0$. In the interest of notational compactness, the time dependence of the matrices will not be shown for the remainder of the chapter. Perfectly known inputs as well as a large class of uncertainties, including parametric uncertainties and neglected dynamics from model reduction, can be absorbed into the above model (Section 3.3.1). To incorporate neglected dynamics, one would naturally have to add some additional states. The zero entries in the last column are due to the fact that the plant state estimate does not affect the plant dynamics. Likewise, the zero entry in the third row is due to the fact that the input noise does not enter into the error definition.

In order to achieve the condition of Eq.(C.2), we define a minmax or game theoretic estimation problem that would minimize an objective with respect to the state estimate $\hat{x}$ in the presence of the worst possible input d and initial state x_0:

$$\min_{\hat{x}} \max_{d,x_0} J_2 \qquad\qquad (C.4)$$

where

$$J_2 \equiv \|e\|^2 + \|\epsilon\|^2 - \gamma^2 (\|d\|^2 + \|x_0 - \hat{x}_0\|^2_{P_0^{-1}})$$

subject to the dynamic constraints of Eq.(C.3).

The next section presents the solution to this game problem, which yields the estimator equation. This solution yields a $J_2 < \|x_0\|^2_{\overline{X}_0}$, provided $\overline{X}_0$ is bounded below by a Riccati matrix. In this case, we achieve $J_1 < \gamma^2$, the robust performance criterion.

C.3   Derivation of the Estimator

The solution to the robust estimation problem in Eq.(C.4) is obtained in two stages, with each stage requiring the solution to a Riccati equation. In the first stage, we bound the terms in the objective function introduced for plant robustness. These terms not affected by the estimate $\hat{x}$. The estimation problem is thereby reduced to one with a simpler form. The second stage consists of solving the transformed problem for the optimal estimate.

Theorem C.2 *The Riccati equation*

$$-\dot{X} \;=\; X(A + BZ^{-1}T'S) + (A + BZ^{-1}T'S)'X$$

$$+\gamma^{-2}XBZ^{-1}B'X + S'S$$

$$X(t_f) \;\; = \;\; 0 \tag{C.5}$$

where

$$Z \;\; \equiv \;\; I - \gamma^{-2}(T'T)$$

has a solution $\ni Z > 0 \; \forall t \in [0, T)$, if and only if

$$d^* \;\; \equiv \;\; \gamma^{-2}Z^{1/2}(B'X + S'T)x \tag{C.6}$$
$$\bar{d} \;\; \equiv \;\; Z^{1/2}d - d^*$$

result in

$$\|\epsilon\|^2 - \gamma^2(\|d\|^2 + \|x_0\|^2_{\overline{X}_0}) < -\gamma^2\|\bar{d}\|^2 \tag{C.7}$$

Remarks: The proof is similar to the discrete case. Note that the matrix X is necessarily positive semidefinite.   ◇

For the second stage, we seek a state estimate $\hat{x}$ such that $J_2 < 0$. But first we must define our estimation problem in terms of $\bar{d}$. Define a new objective function J_3 as:

$$J_3 \;\; \equiv \;\; \|e\|^2 - \gamma^2(\|\bar{d}\|^2 + \|x_0 - \hat{x}_0\|^2_{P_0^{-1}})$$
$$> \;\; J_2 - x_0'X_0x_0 \tag{C.8}$$

and solve the optimization problem of Eq.(C.4), with J_3 replacing J_2, subject to the constraints of Eq.(C.3). To do so, we must express the state and observation equations (C.3) in terms of $\bar{d}$. Substituting $d = Z^{-1/2}(\bar{d} + d^*)$ into these equations, we get

$$\dot{x} \;\; = \;\; \overline{A}x + \overline{B}\,\bar{d} \tag{3.9a}$$
$$y \;\; = \;\; \overline{C}x + \overline{D}\,\bar{d} \tag{3.9b}$$

where

$$\overline{A} \;\; = \;\; A + \gamma^{-2}B(B'X + S'T) \tag{3.10a}$$
$$\overline{B} \;\; = \;\; BZ^{-1/2} \tag{3.10b}$$
$$\overline{C} \;\; = \;\; C + \gamma^{-2}D(B'X + S'T) \tag{3.10c}$$
$$\overline{D} \;\; = \;\; DZ^{-1/2} \tag{3.10d}$$

Note that we have now reduced the estimation problem to a formulation similar to that when the plant dynamics are known, i.e., the continuous analog of the problem of Section 3.2.

The dynamics of the estimator will be assumed to have the same form as those of the Kalman filter:

$$\dot{\hat{x}} = (\overline{A} - K\overline{C})\hat{x} + Ky \tag{3.11}$$

It was shown in [7] that the optimal solution for the game theoretic estimation problem has the above form. The dynamics of the estimation error $\tilde{x} \equiv x - \hat{x}$ can then be represented as

$$
\begin{aligned}
\dot{\tilde{x}} &= (\overline{A} - K\overline{D})\tilde{x} + (\overline{B} - K\overline{C})\overline{d} \\
&= \tilde{A}\tilde{x} + \tilde{B}\overline{d} \\
e &= M\tilde{x}
\end{aligned} \tag{3.12}
$$

The new estimation problem can now be described as:

$$\min_{K} \max_{\overline{d}, \tilde{x}_0} J_3 \quad \text{given } \dot{\tilde{x}} = \tilde{A}\tilde{x} + \tilde{B}\overline{d} \tag{3.13}$$

To solve this problem, we first maximize with respect to $\overline{d}$. If 2λ are the Lagrange multipliers associated with the constraint, then our augmented cost function is:

$$J_3 + \int_0^{t_f} 2\lambda'(\dot{\tilde{x}} - \tilde{A}\tilde{x} - \tilde{B}\overline{d})dt \tag{3.14}$$

Taking the variation and setting it to zero, we get

$$
\begin{aligned}
\overline{d}^* &= -\gamma^{-2}\tilde{B}'\lambda \tag{3.15a} \\
\tilde{x}^*(0) &= -\gamma^{-2}\overline{P}_0\lambda(0) \tag{3.15b}
\end{aligned}
$$

The Hamiltonian equation is:

$$
\begin{bmatrix} \dot{\tilde{x}} \\ \dot{\lambda} \end{bmatrix} = \begin{bmatrix} \tilde{A} & -\gamma^{-2}\tilde{B}\tilde{B}' \\ M'M & -\tilde{A}' \end{bmatrix} \begin{bmatrix} \tilde{x} \\ \lambda \end{bmatrix} \tag{3.16}
$$

with boundary conditions

$$
\begin{aligned}
\tilde{x}(0) &= -\gamma^2\overline{P}_0\lambda(0) \\
\lambda(T) &= 0
\end{aligned}
$$

The corresponding Riccati equation is:

$$
\begin{aligned}
\dot{P} &= \tilde{A}P + P\tilde{A}' + \gamma^{-2}PM'MP + \tilde{B}\tilde{B}' \tag{3.17a} \\
P_0 &= \text{given} \tag{3.17b}
\end{aligned}
$$

The optimal gain K is obtained by substituting the values of the worst case noise and initial condition, $\overline{d}^*$ and $\tilde{x}(0)^*$ respectively of Eqs.(3.15a,3.15b), into the objective function J_3, and optimizing the resulting expression with respect to K. Thus, let $J_4 = \max_{\overline{d},\tilde{x}(0)} J_3$. Apart from a multiple factor independent of K, we have

$$
\begin{aligned}
J_4 &= \int_0^{t_f} \text{trace}\left(PM'MP + \tilde{B}\tilde{B}' + 2\tilde{A}P\right)dt \\
&= \int_0^{t_f} \text{trace}\left(PM'MP + (\overline{B} - K\overline{D})(\overline{B} - K\overline{D})' \right. \\
&\qquad \left. + 2(\overline{A} - K\overline{C})P\right)dt
\end{aligned}
$$

Optimizing with respect to K, we get

$$
K^* = (P\overline{C}' + \overline{B}\,\overline{D}')(\overline{D}\,\overline{D}')^{-1} \tag{3.18}
$$

while the Hessian with respect to K is (apart from a multiple positive factor independent of K):

$$
\overline{D}\,\overline{D}' > 0 \tag{3.19}
$$

which ensures a global minimum since J_4 is quadratic and $\overline{D}\,\overline{D}'$ is assumed to be invertible. The requirement that $\overline{D}\,\overline{D}'$ be invertible implies that all measurements must be noisy. The case where some measurements are perfect is not considered in this derivation.

The estimator dynamics are therefore given by Eq.(3.11), with matrices $\overline{A}$, $\overline{C}$ given by Eq.(3.10a) and Eq.(3.10c) respectively, in terms of the nominal matrices and the Riccati matrix of Eq.(C.5), and with gain K given by Eq.(3.18) in terms of the Riccati matrix of Eq.(3.17a). The next theorem states that the estimator achieves the desired bound of Eq.(C.1).

Theorem C.3 *Assume that for a $\gamma \leq 1$,*

1. *There exists a $X(t)$, necessarily positive semidefinite, satisfying the Riccati Equation (C.5) on $[0, t_f]$ such that $Z(t)$ is positive definite,*

2. *There exists a $\overline{P}(t) > 0$ on the same interval satisfying the Riccati Equation (3.17a) such that $\overline{P}(t) > 0$, and*

3. *$\frac{\gamma^2}{2}\overline{X}_0 > X_0$*

*Then the robust performance condition of Eq.(C.2) is satisfied for $\gamma <$
1.*

Finally, the above theorem, together with the fact that $\bar{D}\bar{D}'$ is positive
definite, shows that the the variational problem solution provides a
game saddle point.

Theorem C.4 $(K^*, (\overline{d}^*, x^*(0)))$ *satisfy:*

$$J_3(K^*, \overline{d}, x(0)) \leq J_3(K^*, \overline{d}^*, x^*(0)) \leq J_3(K, \overline{d}^*, x^*(0)) \qquad (3.20)$$

C.4   Related Work

The steady state continuous-time minmax or H_∞ estimator for nominal
systems was derived by Appleby [4]. Khargonekar [68] and Banavar and
Speyer [7] derived a similar estimator for time-varying finite-horizon
systems with an arbitrary initial condition. Other work is that of DeS-
ouza et al. [21]. The continuous-time steady state μ estimator of [4, 5]
is applicable to a general class of model uncertainties for continuous-
time linear time-invariant systems at steady state only. Specifically,
the first Riccati equation for the estimator in [4, 5] is obtained by spectral factorization. In brief, the robust estimator in this chapter can
accommodate a general class of model uncertainties. Moreover, it is
applicable to time-varying or time-invariant systems, defined over an
infinite or finite horizon, and with an arbitrary initial condition.

APPENDIX D
APPLICATION DATA

This appendix contains relevant numerical data for the two applications discussed in the book. Section D.1 describes the Advanced Research Project Agency's (ARPA) underwater vehicle used as an example in Chapter 5 and 6, while data for the reentry vehicle application is described in Section D.2.

D.1   Underwater Vehicle Application

D.1.1   The Plant

The ARPA unmanned underwater vehicle (UUV) is built in many configurations, depending on its mission, payload, and sensor packages. The two configurations considered here differ in their mass, length and inertia. The nominal configuration has a length of 35.9 feet, a mass of 623 slugs, and a moment of inertia about the y axis of 484,000 slug-sq.ft. The second configuration has a length of 43.9 feet, a mass of 793 slugs, and a moment of inertia about the y axis of 977,700 slug-sq.ft. The vehicle has two vertical and two horizontal fins.

For our example, we use the pitch dynamics in order to detect a failure in the pitch control channel, i.e, in either of the two horizontal fins. The pitch dynamics have three states: pitch angle (deg), pitch rate (deg/sec), and depth rate (ft/sec). When linearized, the continuous-time nominal and perturbed linear models for the longitu-

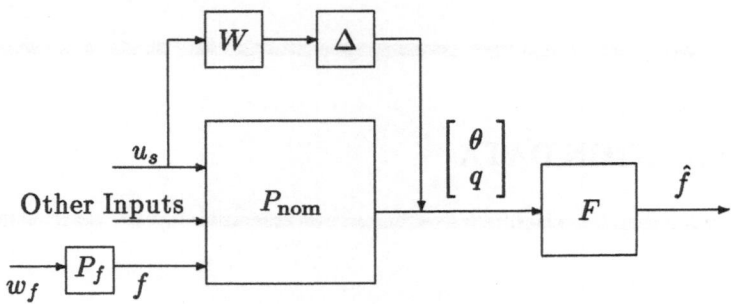

Figure D.1: Plant, uncertainty, and failure model for robust filter design.

dinal dynamics are, respectively

$$A_n = \begin{bmatrix} 0 & 1 & 0 \\ -.0309 & -.5603 & 1.5712 \\ 0 & .0448 & -.1661 \end{bmatrix}, \quad B_n = \begin{bmatrix} 0 & 0 \\ .04 & -.1866 \\ 0 & -.0140 \end{bmatrix}$$

$$A_p = \begin{bmatrix} 0 & 1 & 0 \\ -.0056 & -.4490 & 1.0301 \\ 0 & .045 & -.1299 \end{bmatrix}, \quad B_p = \begin{bmatrix} 0 & 0 \\ .04 & -.0323 \\ 0 & -.0031 \end{bmatrix}$$

$$\tag{4.1}$$

The first column of the B matrices is for the process noise input, while the second column is for inputs from the pitch control channel (stern plane commands). The process noise enters through the pitch rate state, and has a unit variance. Both vehicles use the same sensors. For the pitch dynamics, the pitch angle and pitch rate measurements are of interest

$$C = \begin{bmatrix} 1 & 0 & 0 \\ 0 & 1 & 0 \end{bmatrix} \tag{4.2}$$

The sensor noise has a standard deviation of 1 deg for the pitch angle, and .25 deg/sec for the the pitch rate.

D.1.2   Description of Robust Filter Design

A continuous-time, steady-state formulation is used. Figure D.1 shows the representation for which the robust filter is designed. The design plant P_{nom} is the nominal UUV dynamics. The uncertainty $W\Delta$ is constructed from the difference between the nominal and perturbed configurations. Since the perturbed configuration is unlikely to be the worst case plant at that distance from the nominal, we avoid conservatism by scaling the uncertainty bound with a factor of .5.

A key part in the design of a robust estimator is the uncertainty model. Here, we assume that the uncertainty enters the system additively, as shown in Figure D.1. The input to the uncertainty is the stern plane command, u_s, and the output is added to the the pitch angle, θ and the pitch rate, q. The weight W and the uncertainty Δ have the form:

$$W \;=\; \begin{bmatrix} W_1 \\ W_2 \end{bmatrix}$$

$$\Delta \;=\; \begin{bmatrix} \Delta_1 & 0 \\ 0 & \Delta_2 \end{bmatrix}$$

Let $T^{\text{nom}}_{\theta u_s}$ and $T^{\text{pert}}_{\theta u_s}$ be the transfer functions between u_s and θ for the nominal and the perturbed design configurations, respectively. Similarly, let $T^{\text{nom}}_{q u_s}$ and $T^{\text{pert}}_{q u_s}$ designate the transfer functions between u_s and q for the nominal and the perturbed design configurations, respectively. Then, W_1 and W_2 are given by

$$W_1 \;=\; T^{\text{pert}}_{\theta u_s} - T^{\text{nom}}_{\theta u_s}$$

$$W_2 \;=\; T^{\text{pert}}_{q u_s} - T^{\text{nom}}_{q u_s}$$

The Bode plots of the two transfer functions W_1 and W_2 are shown in Figure D.2. Finally, the shaping filter for the failure has a bandwidth of $-3. \times 10^{-6}$ and an amplitude of of $4.188. \times 10^6$.

D.2   Reentry Vehicle Application

D.2.1   The Plant

The models considered are linearizations of the the space shuttle Orbiter plant at various Mach Numbers, angles of attack, and constant altitudes [124]. They are originally given in continuous-time, and are discretized using a sampling period of $T = 0.005$ seconds. The six states are as follows: bank angle (deg), angle of attack (deg), sideslip angle (deg), bank rate (deg/sec), time rate of change angle of attack (deg/sec), and sideslip rate (deg/sec). The inputs are aileron deflection (deg), elevator deflection (deg), rudder deflection (deg), and the multiple jet command (lb). The measurements are the bank and sideslip

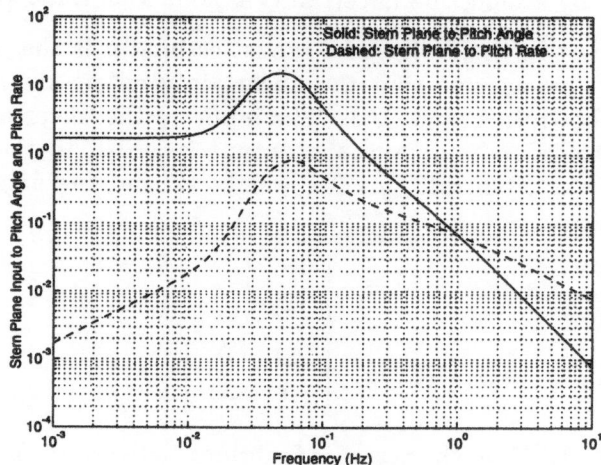

Figure D.2: Magnitude of frequency response bound for uncertainty model.

angle rates (*deg/sec*), and the rate of change of the angle of attack. The filters are the subjects of Agustin's thesis [2].

```
Nominal plant, M = 7.5
Angle of attack = 35 degrees
Altitude = 150 kft

Anom =
   1.0000e+00              0   -4.2794e-05    4.9998e-03             0   -7.1323e-08
            0     1.0000e+00              0             0    4.9998e-03             0
            0              0    9.9998e-01    2.7757e-08             0    4.9998e-03
            0              0   -1.7117e-02    9.9994e-01             0   -4.2793e-05
            0    -1.5932e-03              0             0    9.9991e-01             0
            0              0   -8.6480e-03    1.1103e-05             0    9.9990e-01

Bnom =
   1.6302e-05              0    3.6681e-07   -2.0971e-09
            0    -4.5522e-06              0   -1.1245e-10
   1.0810e-05              0    2.7025e-07   -6.4574e-09
   6.5209e-03              0    1.4672e-04   -8.3882e-07
            0    -1.8208e-03              0   -4.4980e-08
   4.3240e-03              0    1.0810e-04   -2.5829e-06

Gnom =
   8.1513e-07              0   -9.6379e-14   -2.0971e-10  0  0  0
            0    -2.2761e-07              0   -1.1245e-11  0  0  0
   3.0169e-12              0    1.3513e-08   -6.4574e-10  0  0  0
   3.2605e-04              0   -7.7103e-11   -8.3882e-08  0  0  0
```

```
          0  -9.1042e-05                  0  -4.4980e-09  0  0  0
    1.8101e-09            0   5.4050e-06  -2.5829e-07  0  0  0
```

Cnom =
```
    0  0  0  8.1915e-01  0                    5.7358e-01
    0  0  0  0                  1.0000e+00  0
    0  0  0  5.7358e-01  0                   -8.1915e-01
```

Dnom =
```
    0  0  0  0  7.0711e-03  0                    0
    0  0  0  0  0                  7.0711e-03  0
    0  0  0  0  0                  0                    7.0711e-03
```

Perturbed plant, M = 8.8
Angle of attack = 38 degrees
Altitude = 158 kft

Apert =
```
    1.0000e+00               0  -4.4245e-05   4.9999e-03                0  -7.3741e-08
            0   1.0000e+00               0                0   4.9998e-03                0
            0               0   9.9997e-01   3.3098e-08                0   4.9998e-03
            0               0  -1.7698e-02   9.9995e-01                0  -4.4244e-05
            0  -1.6114e-03               0                0   9.9991e-01                0
            0               0  -1.0817e-02   1.3239e-05                0   9.9991e-01
```

Bpert =
```
    1.5390e-05               0  -1.8813e-15  -1.7563e-09
            0  -5.9087e-06               0  -1.1245e-10
    1.4287e-05               0   2.5512e-10  -6.5583e-09
    6.1557e-03               0  -1.5051e-12  -7.0249e-07
            0  -2.3635e-03               0  -4.4980e-08
    5.7146e-03               0   1.0205e-07  -2.6233e-06
```

Cpert =
```
    0  0  0  8.1915e-01  0                    5.7358e-01
    0  0  0  0                  1.0000e+00  0
    0  0  0  5.7358e-01  0                   -8.1915e-01
```

Dpert =
```
    0  0  0  0  7.0711e-03  0                    0
    0  0  0  0  0                  7.0711e-03  0
    0  0  0  0  0                  0                    7.0711e-03
```

D.2.2   The Filters

The following matrices are for some of the various filters used. First,
the Kalman filter design based on the nominal plant. This filter is used
in the attitude determination example of Section 3.6.2, used also in the

FDI architecture of Section 6.3.2. Next, matrices for the overdesigned Kalman filter discussed in Section 3.6.2 are given. Then the robust attitude determination filter matrices are listed.

Kalman filter ($F2$)

k =

```
   3.9961e-03    5.5646e-13    2.7983e-03
  -5.4967e-10   -2.5281e-04    1.9918e-10
  -1.4237e-04    4.4593e-10   -9.2920e-05
   3.6938e-02    5.8337e-12    2.5812e-02
   8.3531e-10    1.2740e-02   -1.1828e-09
   5.0175e-04    1.4480e-09   -6.4103e-04
```

a - k*c =

```
  1.0000e+00            0   -4.2794e-05    1.2145e-04   -5.5646e-13    9.2290e-08
           0   1.0000e+00            0    3.3602e-10    5.2526e-03    4.7843e-10
           0            0    9.9998e-01    1.6995e-04   -4.4593e-10    5.0053e-03
           0            0   -1.7117e-02    9.5487e-01   -5.8337e-12   -8.6121e-05
           0   -1.5932e-03            0   -5.8337e-12    9.8717e-01   -1.4480e-09
           0            0   -8.6480e-03   -3.2225e-05   -1.4480e-09    9.9909e-01
```

Overdesigned Kalman filter ($F2$)

k =

```
   2.1360e-03   -2.6066e-12    1.4958e-03
  -1.3512e-08   -1.4366e-01    1.2634e-08
  -2.1939e-04   -3.3101e-09    2.4988e-04
   7.8387e-01    9.8169e-10    5.4887e-01
   5.1540e-08    7.0248e-01   -7.1895e-08
   4.2234e-02    8.8455e-08   -6.0306e-02
```

a - k*c =

```
  1.0000e+00            0   -4.2794e-05    2.3922e-03    2.6066e-12    1.1796e-08
           0   1.0000e+00            0    3.8219e-09    1.4866e-01    1.8100e-08
           0            0    9.9998e-01    3.6417e-05    3.3101e-09    5.3303e-03
           0            0   -1.7117e-02    4.3007e-02   -9.8169e-10   -4.8724e-05
           0   -1.5932e-03            0   -9.8169e-10    2.9743e-01   -8.8455e-08
           0            0   -8.6480e-03    5.1711e-06   -8.8455e-08    9.2628e-01
```

Robust Filter $F1$ (One of the steady state filters used for thrust estimation (from end of simulation, no resetting. The angle of attack and its rate is not used in this design, and only the 1st and 3rd measurements are used.)

k=

```
   2.5704e-03    1.5067e-03
  -6.3277e-02   -8.3991e-02
   5.6749e-01    5.0357e-01
   3.4184e-01    8.7696e-02
  -3.6359e+01    2.4004e+02
```

a - k*c =

```
   1.0000e+00   -6.0656e-05    2.0297e-03   -2.4019e-04   -4.7942e-09
            0    9.9997e-01    1.0002e-01   -2.7512e-02   -8.2844e-09
            0   -2.4262e-02    2.4612e-01    8.6984e-02   -1.9178e-06
            0   -1.3380e-02   -3.3037e-01    8.7567e-01   -3.2932e-06
            0    4.5797e-05   -1.0791e+02    2.1751e+02    1.0000e+00
```

Robust Filter $F2$ (for navigation and aileron failure detection)

k =

```
   3.5759e-03   -3.7558e-15    2.7492e-03
   1.6122e-13   -6.6958e-02   -8.4897e-13
  -2.2703e-03   -1.5850e-14   -6.7268e-03
   2.0845e-01    1.4705e-12    4.7745e-02
   3.3133e-12    3.3929e-01   -2.1680e-12
   3.3714e-01    3.6763e-12   -3.4123e-01
```

a - k*c =

```
   1.0000e+00   -4.7778e-19   -4.2753e-05    4.9368e-04    3.7832e-15    2.0090e-04
            0    1.0000e+00   -2.9928e-17    3.5492e-13    7.1960e-02   -7.8786e-13
            0   -1.4752e-18    9.9998e-01    5.7183e-03    1.5935e-14    7.9164e-04
            0   -1.9111e-16   -1.7101e-02    8.0181e-01   -1.4596e-12   -8.0485e-02
            0   -1.5934e-03   -1.1971e-14   -1.4600e-12    6.6062e-01   -3.6431e-12
            0   -5.9005e-16   -8.6393e-03   -8.0431e-02   -3.6428e-12    5.2702e-01
```

BIBLIOGRAPHY

[1] Adams, M., *Linear Estimation of Boundary Value Stochastic Processes*, Ph.D. Thesis, Department of Aeronautics and Astronautics, M.I.T., and Draper Laboratory Report No. CSDL-T- , 1983.

[2] Agustin, R., *Robust Estimation and Failure Detection for Reentry Vehicle Attitude Control Systems*, Master of Science Thesis, Department of Mechanical Engineering, M.I.T., and Draper Laboratory Report No. CSDL-T-1301, June, 1998.

[3] Anderson, B.D.O., and Moore, J., *Optimal Filtering*, Prentice Hall, Englewood Cliffs, New Jersey, 1979.

[4] Appleby, B., *Robust Estimator Design using the H_∞ Norm and μ Synthesis*, Ph.D. Thesis, Department of Aeronautics and Astronautics, MIT, and Draper Laboratory Report No. CSDL-T-1065, 1990.

[5] Appleby, B., Dowdle, J., and VanderVelde, W., 'Robust Estimator Design using μ Synthesis," *Proceedings of the IEEE Conference on Decision and Control*, pp. 640-645, December 1991.

[6] Astrom, K,, and Wittenmark, J., *Adaptive Control*, Second Edition, Addison Wesley, 1995.

[7] Banavar, R. and Speyer, J., "A Linear Quadratic Game Approach to Estimation and Smoothing," *Proceedings of the American Control Conference*, pp. 2818-2822, June 1991.

[8] Bartlett, A.C., Hollot, C.V., and Lin, H., "Root Locations of an Entire Polytope of Polynomials: It Suffices to Check the Edges," in *Mathematics of Control, Signals, and Systems*, Springer Verlag, 1988.

[9] Basar, T., and Olsder, G.L., *Dynamic Non-Cooperative Game Theory*, Math. Sci. Eng., vol.160, Academic Press, New York, 1982.

[10] Basar, T., and Bernhard, P., H_∞ *Optimal Control and Related Minimax Design Problems*, Birkhauser, second edition, 1995.

[11] Basar, T., "A Dynamic Games Approach to Controller Design: Disturbance Rejection in Discrete-Time," *Proceedings of the IEEE Conference on Decision and Control*, pp. 407-414, December 1989.

[12] Basseville, M., "Detecting Changes in Signals and Systems-A Survey," *Automatica*, vol. 24, pp. 309-326, 1988.

[13] Basseville, M., and Nikiforov, I., *Detection of Abrupt Changes, Theory and Application*, Prentice Hall, 1993.

[14] Battin, R.H., "A Statistical Optimizing Navigation Procedure for Space Flight," *Journal of the American Rocket Society*, vol.32, pp. 1681-1692, 1962.

[15] Battin, R.H., *Astronautical Guidance*, McGraw Hill, New York, 1964.

[16] Beard, R., *Failure Accommodation in Linear Systems Through Self-Reorganization*, Rept. MVT-71-1, Man Vehicle Laboratory, MIT, 1971.

[17] Bensoussan, A., and Van Schuppen, J.H., "Optimal Control of Partially Observable Stochastic Systems with an Exponential-of-Integral Performance Index," *SIAM Journal of Control and Optimization*, vol. 23, no.4, July 1985.

[18] Bryson, R., and Ho, Y., *Applied Optimal Control*, Hemisphere Publishing Co., 1969.

[19] Dahleh, M.A., and Pearson, J.B., "ℓ_1-optimal feedback controllers for MIMO discrete-time systems", *IEEE Transactions on Automatic Control*, vol. 32, No. 4, pp.314-322, April 1987.

[20] Desoer, C., and Vidyasagar, M., *Feedback Systems: Input-Output Properties*, Academic Press, New York, 1975.

[21] De Souza, C., Shaked, U., and Fu, M. "Robust H_∞ Filtering with Parametric Uncertainty and Deterministic Input Signals," *Proceedings of the IEEE Conference on Decision and Control*, pp. 2305-2310, December 1992.

[22] Deyst, J., "A Derivation of the Optimum Continuous Linear Estimator for Systems with Correlated Measurement Noise," *AIAA Journal*, vol. 7, no.11, pp. 2116-2119, November, 1969.

[23] Dorato, P., ed., *Robust Control*, IEEE Press, New York, 1987.

[24] Doyle, J., Glover, K., Khargonekar, P., and Francis, B., "State-Space Solutions to the H_2 and H_∞ Control Problems," *IEEE Transactions on Automatic Control*, vol.34, no.8, pp.831-847, August 1987.

[25] Doyle, J., "Analysis of Feedback Systems with Structured Uncertainties," *IEE Proceedings*, vol. 129, pp. 242-250, 1982.

[26] Doyle, J., "Structured Uncertainty in Control System Design," *Proceedings of the IEEE Conference on Decision and Control*, pp. 260-265, December 1985.

[27] Edelmayer, A., Bokor, J., and Kevizsky, L., "An H_∞ Filtering Approach to Robust Detection of Failures in Dynamic Systems," to be published in *Proceedings of the IEEE Conference on Decision and Control*, December 1994.

[28] Emami-Naeini, A., Akhter, M., and Rock, S., "Robust Detection, Isolation, and Accomodation for Sensor Failures," *Proceedings of the American Control Conference*, pp. 1052-1059, June 1986.

[29] Emami-Naeini, A., Akhter, M., and Rock, S., "Effect of Model Uncertatinty on Failure Detection," *IEEE Transactions on Automatic Control*, vol.33, no.12, pp.1106-1115, December 1988.

[30] Farrell, J., Berger, T., and B. Appleby, "Detection and Accommodation of Unanticipated Faults via Learning Techniques," *IEEE Control Systems Magazine: Special Issue on Intelligent Control*, pp.40-49, June 1993.

[31] Francis, B.A., Helton, J.W., and Zames, G., "H_∞-Optimal Feedback Controllers for Linear Multivariable Systems," *IEEE Transactions on Automatic Control*, vol.29, no.10, pp. 818-899, October, 1984.

[32] Frank, P., and Ding, X., "Frequency Domain Approach to Optimally Robust Residual Generation and Evaluation for Model-based Fault Diagnosis," *Automatica*, Vol. 30, No. 5, 1994, pp. 789-804.

[33] Frank, P.M., and Wuennenberg, J., "Robust Fault Diagnosis using Unknown Input Observer Schemes," in *Fault Diagnosis in Dynamic Systems: Theory and Application*, Patton, R.J., Frank, P.M., and Clark, R.N., eds., Prentice-Hall, N.J., 1989.

[34] Fraser, D., and Potter, J., "The Optimum Linear Smoother as a Combination of Two Optimum Linear Filters," *IEEE Tansactions on Automatic Control*, vol.7, no.8, pp.387-390, August 1969.

[35] Frank, P.M., "Fault Diagnosis in Dynamic Systems Using Analytical and Knowledge Based Redundancy," *Automatica*, vol.26, pp.459-474, 1990.

[36] Gertler, J., and Singer, D., "A New Structural Framework for Parity Equation Based Failure Detection and Isolation," *Automatica*, vol.26, pp.381-388, 1990.

[37] Gertler, J., "Survey of Model-Based Failure Detection and Isolation in Complex Plants," *Control System Magazine*, pp. 3-11, December 1991.

[38] Glover, K., "All Optimal Hankel-Norm Approximations of Linear Multivariable Systems and their L_∞-Error Bounds," *IEEE Transactions on Automatic Control*, vol.21, pp.319-338, 1976.

[39] Green, M., and Limebeer, D.J., *Linear Robust Control*, Prentice Hall, Englewood Cliffs, N.J., 1994.

[40] Goodwin, G., and Sin, K.S., *Adaptive Filtering, Prediction, and Control*. Information and Systems Science Series, Prentice Hall, Englewood Cliffs, NJ.

[41] Graham, W., and McLaughlin, D., "Stochastic Analysis of Nonstationary Subsurface Solute Transport, 2. Conditional Moments," *Water Resources Research*, vol. 25, no. 11, pp. 2331-2355, November 1989.

[42] Grenander, U., *Abstract Inference*, Wiley, New York, 1981.

[43] Grimble, M., "H_∞ Design of Optimal Linear Filters," in Byrnes, C.I., Martin, C.F., and Sacks, R.E., editors, *Linear Circuits, Systems, and Signal Processing; Theory and Application*, pp.533-540. North Holland, Amsterdam, 1988. Proceedings of MTNS, Phoenix, Arizona, 1987.

[44] Grimble, M., Ho, R., and Elsayed, A., H_∞ Robust Linear Estimators. In *IFAC symposium*, 1989.

[45] Grimble, M., "H_∞ Fixed-Lag Smoothing Filter for Scalar Systems," *IEEE Transanctions On Signal Processing*, vol.19, no.9, pp.1955-1963, September 1991.

[46] Guan, Y., and Saif, M., "A Novel Approach to the Design of Unknown Input Observers," *IEEE Transactions on Automatic Control*, vol.36, pp.3632-635, 1991.

[47] Hall, S., *A Failure Detection Algorithm for Linear Dynamic Systems*, Ph.D. Thesis, Department of Aeronautics and Astronautics, MIT, 1985.

[48] Herring, T.A., Davis, J.L., Shapiro, I.I., Geodesy by Radio Interferometry: The Application of Kalman Filtering to the Analysis of Very Long Baseline Interferometry Data, *Journal of Geophysics Research*, v.95, no.B8, pp. 12,561-12,581, 1990.

[49] Hassibi, B., Sayed, A. H., and Kailath, T., "Recursive linear estimation in Krein spaces - part I: Theory," *Proc. IEEE Conference on Decision and Control*, pp. 3489-3495, December 1993.

[50] Hassibi, B., Sayed, A. H., and Kailath, T., "Recursive linear estimation in Krein spaces - part II: Applications," *Proc. IEEE Conference on Decision and Control*, pp. 3495-3501, December, 1993.

[51] Ho, N., Lozano, P., Mangoubi, R., and Martinez-Sanchez, M., "Failure Detection and Isolation for the Space Shuttle Main Engine," *Proceedings of the joint AIAA/ASME/AIChE Conference on Propulsion*, Seattle, Wash., July, 1997.

[52] Horak, D., "Failure Detection in Dynamic Systems with modeling Errors," *Journal of Guidance and Control*, pp. 508-516, Nov.-Dec., 1988.

[53] How, J., *Robust Control Design with Real Parameter Uncertainty Using Absolute Stability Theory*, Ph.D. Thesis, Department of Aeronautics and Astronautics, MIT, 1993.

[54] Huber, P., *Robust Statistics*, Wiley, New York, 1981.

[55] Huber, P., Written communication.

[56] Isermann, R., "Process Fault Detection Based on Modeling and Estimation" *Automatica*, vol. 20, pp. 387-404, 1984.

[57] Jacobson, D., "Optimal Stochastic Linear Systems with Exponential Performance Criteria and their Relation to Deterministic Differential Games," *IEEE Transactions on Automatic Control*, vol.18, no.2, pp. 124-131, 1973.

[58] Jacquemont, C., *Aircraft Attitude Determination Using Robust Estimation*, Master of Science Thesis, M.I.T., Department of Aeronautics and Astronautics, August, 1997.

[59] Jones, R., *Failure Detection in Linear Systems*, Ph.D. Thesis, Department of Aero/Astro Eng., MIT, 1973.

[60] Kailath, T., "An Innovation Approach to Least-Squares Estimation, Part I: Linear Filtering in Additive White Noise," *IEEE Transactions on Automatic Control*, vol.13, no.6, pp. 646-655, 1968.

[61] Kailath, T., "A View of Three Decades of Linear Filtering Theory," *IEEE Transactions on Information Theory*, vol.20, no.2, pp. 146-181, 1974.

[62] Kailath, T., *Lectures on Wiener and Kalman Filtering*, CISM Courses and Lectures, no.140, New York: Springer-Verlag, 1981.

[63] Kalman, R.E., "A New Approach to Linear Filtering and Prediction Problems," *Transactiaon of the ASME Journal of Basic Engineering*, vol. 82D, pp. 35-45, March 1960.

[64] Kalman, R.E., and Bucy, R. "New Results in Linear Filtering and Prediction," *Transactiaon of the ASME Journal of Basic Engineering*, vol.83D, pp. 95-108, March 1961.

[65] Kalman, R.E. "New Methods of Wiener Filtering Theory," in *Proc. 1st Symp. Engineering Applications of Random Functions Theory and Probability*, J.L. Bogdanoff and F. Kozin, Eds., Wiley, New York, pp. 270-388, 1963.

[66] Kassam, S.A., and Poor, H.V., "Robust Techniques for Signal Processing: A survey," *Proceedings of the IEEE*, vol. 73, no.3, pp. 433-480, March 1985.

[67] Kharitonov, V.L., "Assymptotic Stability of an Equilibruim Position of a Family of Systems of Linear Differential Equations," *Differential Equations*, vol. 14, pp. 1483-1485, 1979.

[68] Khargonekar, P.P., and Nagpal, K.M., "Filtering and Smoothing in an H_∞ Setting," *IEEE Transactions on Automatic Control*, vol. 36, no. 2, pp. 152-166, December 1989.

[69] Koifman, M., and Merhav, S., "Autonomously Aided Strapdown Attitude Reference Systems," *Journal of Guidance, Dynamics, and Control*, Vol. 14, No.6, Nov.-Dec. 1991, pp.1164-1172.

[70] Kumar, P., and Van Schuppen, J., "On the Optimal Control of Stochastic Systems with and Exponential-of-Integral Peformance Index," *Journal of Mathematical Analysis and Applications*, vol.80, pp. 312-332, 1981.

[71] Kumar, and Varaiya, *Stochastic Systems: Estimation, Identification, and Adaptive Control*, Prentice Hall, Englewood Cliffs, N.J., 1986.

[72] Kuo, B.C., *Automatic Control Systems*, fourth edition, Prentice-Hall, N.J.,1982.

[73] Laning, H., and Battin, R., *Random Processes in Automatic Control*, McGraw-Hill, New York, 1958.

[74] Lewis, F., *Optimal Estimation*, John Wiley and Sons, 1986.

[75] Limebeer, D., Anderson, B., Khargonekar, P., and Green, M. "A Game Theoretic Approach to H_∞ Control for Time-Varying Systems," *SIAM Journal of Control and Optimization*, vol. 30, no.2, pp. 262-283, March 1992.

[76] Lou, X., Willsky, A., and Verghese, G., "Optimally Robust Redundancy Relations for Failure Detection in Uncertain Systems," *Automatica*, vol. 22, pp. 333-344, 1986.

[77] Ljung, L., *System Identification - Theory for the User*, Prentice Hall, 1987.

[78] Luenberger, D. "An Introduction to Observers," *IEEE Transactions on Automatic Control*, vol. 16, no.6, pp. 596-602, 1971.

[79] Mangoubi, R., Appleby, B., and Farrell, J., "Robust Estimation in Failure Detection," *Proceedings of the IEEE Conference on Decision and Control*, pp. 2317-2322, December, 1993.

[80] Mangoubi, R., Appleby, B., and Verghese, G. "Robust Estimation for Discrete-Time Linear Systems," *ACC Conference on Decision and Control*, pp. 656-661, May 1994.

[81] Mansour, M., Balemi, S., and Truol, W., eds., *Robustness of Dynamic Systems with Parameter Uncertainties*, Birkhauser Verlag, Basel, 1992.

[82] Matthews, M., *Fault Slip in Space and Time*, Ph.D. Thesis, Stanford University, 1991.

[83] Maybeck, x. *Stochastic Models, Estimation, and Control, Volume 1*, Academic Press, New York, 1979.

[84] Merhav, S., *Aerospace Sensors and Their Applications,*, Springer Verlag, 1996.

[85] Nash, R.A., Kaisper, J.F., Crawford, B.S., and Levine, S.A., "Application of Optimal Smoothing to the Testing and Evaluation of Inertial Navigation Systems and Components," *IEEE Transactions on Automatic Control*, Vol. AC-16, No.6, December, 1971, pp.806-816.

[86] Patton, R.J., and Chen, J., "A Review of Parity Space Approaches to Fault Diagnosis," *Proc. SAFEPROCESS'91*, Baden-Baden, Germany, pp.239-256, 1991.

[87] Patton, R.J., Frank, P.M., and Clark, R.N., eds., *Fault Diagnosis in Dynamic Systems: Theory and Application*, Prentice-Hall, N.J., 1989.

[88] Poor, V., *An Introduction to Signal Detection and Estimation*, Springer Verlag, 1988.

[89] Rauch, H.E., "Optimum Estimation of Satellite Trajectories including Random Fluctuations in Drag," *AIAA Journal*, Vol. 3, No. 4, April 1965, pp.717-722.

[90] Ribbens, W.B., and Riggins, R.,N., "The Distinction Between Actuator Failures and Plant Dynamic Changes," *IFAC/IMACS Symposuim on Fault Detection Supervision and Safety for Technical Processes - Safe Process*, Baden-Baden, Germany, 1991.

[91] Riggins, R.N., *The Distinction Between Actuator Failures and Plant Dynamics Changes*, Ph.D. Dissertation, Department of Electrical Engineering and Computer Science, University of Michigan, 1990.

[92] Schick, I., *Robust Recursive Estimation of the State of a Discrete-Time Stochastic Linear Dynamic System in the Presence of Heavy-Tailed Observation Noise*, Ph.D. Thesis, Department of Mathematics, MIT, 1989.

[93] Semyonov, A.V., Vladimirov, I.G., and Kurdjukov, A.P., "A Stochastic Approach to H_∞ Optimization," to be published in the *Proceedings of the 33rd IEEE Conference on Decision and Control*, December 1994.

[94] Siljak, D.D., "Parameter Space Methods for Robust Control Design: A Guided Tour," *IEEE Transactions on Automatic Control*, vol.34, pp. 674-688, 1989.

[95] Slepian, D., "Linear Least-Squares Filtering of Distorted Images," *Journal of the Optical Society of America*, vol. 57, pp. 918-922, July 1967.

[96] Smith, R., and Dahleh, M., eds., *The Modeling of Uncertainty in Control Systems*, Lecture Notes in Control and Information Sciences, no. 192, Springer-Verlag, 1992.

[97] Sorenson, H. W., "Least Square Estimation: from Gauss to Kalman," *IEEE Spectrum*, vol. 7, no. 7, pp.63-38, July 1970.

[98] Sorenson, H.W., ed. *Kalman Filtering: Theory and Application*, IEEE Press, New York, 1985.

[99] Speyer, J., Chih-hai, F., and Banavar, R. "Optimal Stochastic Estimation with Exponential Cost Criteria," *Proceedings of the IEEE Conference on Decision and Control*, pp. 2293-2298, December 1992.

[100] Speyer, J., Deyst, J., and Jacobson, D., "Optimization of Stochastic Linear Systems with Additive Measurement and Process Noise Using Exponential Performance Criteria," *IEEE Transactions on Automatic Control*, vol.19, no.4, pp. 358-366, 1974.

[101] Stengel, R., and Ryan, L., "Stochastic Robustness of Linear Control Systems," *Proceedings of the 1989 Conference on Information Sciences and Systems*, Johns Hopkins University, Baltimore, March 1989.

[102] Strang, G., *Introduction to Applied Mathematics*, Wellesely-Cambridge Press, Wellesely, Mass., 1986.

[103] Strang, G., *Linear Algebra and Its Applications*, second edition, Academic Press, 1980.

[104] Stoorvogel, A., "The Discrete Time H_∞ Control Problem with Measurement Feedback," SIAM Journal of Control and Optimization, Vol. 30, No.1, pp.182-102, January, 1992.

[105] Swerling, P., "First-Order Error Propagation in a Stagewise Smoothing Procedure for Satellite Observations," *Journal of Astraunautical Sciences*, vol. 6, pp. 46-52, 1959.

[106] Tadmor, G., "H_∞ in the Time Domain: The Standard Four Block Problem," *Mathematics of Control Systems and Signal Processing*, Vol. 3, pp. 301-324, 1990.

[107] Tsui, C.C., " A General Failure Detection, Isolation, and Accomodation System with Model Uncertainty and Measurement Noise," *IEEE Transactions on Automatic Control*, vol.39, no.11, pp. 2318-2321, November 1994.

[108] Van Trees, H., *Detection, Estimation, and Modulation Theory*, Wiley, 1968.

[109] Voulgaris, P., "On Optimal ℓ_∞ to ℓ_∞ Filtering," *Automatica*, vol.31, no. 3, pp.489-495, 1995.

[110] Whittle, P., *Prediction and Regulation*, Van Nostrand Reinhold, New York, 1963.

[111] Whittle, P., "Risk-Sensitive Linear/Quadratic/Gaussian Control," *Advances in Applied Probability*, vol.13, pp. 764-777, 1981.

[112] Whittle, P., *Risk Sensitive Optimal Control*, Wiley, 1990.

[113] Wiener, N., *Extrapolation, Interpolation, and Smoothing of Stationary Time Series, with Engineering Applications*, Technology Press and Wiley, 1949.

[114] Willsky, A. and Jones, H., "A Generalized Likelihood Ratio Approach to the Detection and Estimation of Jumps in Linear Systems" *IEEE Transactions on Automatic Control*, vol. 21, pp. 108-112, February 1976.

[115] Willsky, A., "A Survey of Design Methods for Failure Detection in Dynamic Systems," *Automatica*, vol. 12, pp. 601-611, 1976.

[116] Willsky, A., "Detection of Abrupt Changes in Dynamic Systems," in *Detection of Abrupt Changes in Signals and Dynamical Systems*, Basseville, M., and Benvveniste, A., eds. Lecture Notes in Control and Information Sciences, LNCIS 77, Springer, New York, pp.27-49.

[117] Wilson, D. "Extended Optimality Properties of the Linear Quadratic Regulator and Stationary Kalman Filter," *IEEE Transactions on Automatic Control*, vol. 35, pp. 583-585, May 1990.

[118] Woods, J. W., and Radewan, C. H. "Kalman Filtering in Two Dimensions," *IEEE Transactions on Information Theory*, vol.23, pp. 473-482, July 1977.

[119] Xie, L., De Souza, C., and Fu, M. "H_∞ Estimation for Discrete-Time Linear Uncertain Systems," *International Journal of Robust and Nonlinear Control*, vol. 1, pp. 111-123, December 1992.

[120] Yaesh, I. and Shaked, U., "Game Theory Approach to Optimal Linear Estimation in the Minimum H_∞-Norm Sense," *Proceedings of the IEEE Conference on Decision and Control*, pp. 421-425, December 1989.

[121] Yaesh, I. and Shaked, U., "A Transfer Function Approach to the Problems of Discrete-Time Systems: H_∞-Optimal Linear Control and Filtering," *IEEE Transactions on Automatic Control*, vol.30, no.11, pp. 1264-1271, November 1991.

[122] Yaesh, I. and Shaked, U., "Minimum H_∞-norm regulation of linear discrete-time systems and its relation to linear quadratic discrete games," *Proceedings of the IEEE Conference on Decision and Control*, pp. 942-947, December 1989.

[123] Young, P. *Robustness with Parametric and Dynamic Uncertainty*, Ph.D. thesis, California Institute of Technology, Pasadena, Calif., 1993.

[124] Zacharias, G.L., *A Digital Autopilot for the Space Shuttle Vehicle*, Master of Science Thesis, Department of Aeronautics and Astronautics, Massachusetts Institute of Technology, Cambridge, MA, February 1974.

[125] Zames, G., "Feedback and Optimal Sensitivity: Model Reference Transformations, Multiplicative Seminorms, and Approximate Inverses," *IEEE Transactions on Automatic Control*, vol.26, no.2, pp. 301-320, April 1991.

INDEX